The Best American Science and Nature Writing 2003

The Best American Science and Nature Writing 2003

Edited and with an Introduction by Richard Dawkins

Tim Folger, Series Editor

HOUGHTON MIFFLIN COMPANY

BOSTON · NEW YORK 2003

ISSN 1530-1508
ISBN 0-618-17891-0
ISBN 0-618-17892-9 (pbk.)

Printed in the United States of America

MP 10 9 8 7 6 5 4 3 2 1

Contents

Foreword

FOR A FEW CENTURIES NOW a remarkable story has been unfolding. Not everyone likes the story — it shatters myths of every
sort, questions all beliefs, and to be honest, the plot is not always
easy to follow. Worst of all — or best of all, depending on your
point of view — no one knows how or if the story will end, even
though the narrative's pace seems ever to accelerate. The story, of
course, is the one that science tells us about the universe, and of
the emergence on a small, watery speck of a world of an unusually
curious and fractious bipedal species.

Like children getting a first inkling of the wide and intimidating
horizons outside their front door, we *Homo sapiens* are just beginning to understand our true and ever-so-tentative place in the cosmos. For a wrenching sense of perspective on the brevity of our
species' reign to date, Timothy Ferris, one of the contributors to
this volume and a passionate lifelong amateur astronomer, recom-
mends a meditation on the Andromeda galaxy. Andromeda is the
most distant object that can be seen with the naked eye — it's
about 3 million light-years from Earth, visible as a small, fuzzy
patch in the autumn sky. Now, 3 million years ago — which is when
the light from Andromeda that reaches your eyes on any given
night began its journey — is about when the first hominids appeared. While that is certainly worth pondering, Ferris really wants
you to consider a somewhat subtler experience.

Andromeda, remember, is a galaxy like our own Milky Way, only half again as large, and contains about 200 billion stars, herded by gravity into a spiral disk some 150,000 light-years wide. When you look at that island of stars, the light from its far and near edges hits your retina at the same time. But, as Ferris points out, the light from the more distant edge had to travel 150,000 light-years farther — the width of Andromeda — than the light from the galaxy's nearer shore. In a single glimpse of Andromeda, then, your eyes capture light that encompasses a span of 150,000 years, which is roughly equal to the length of time that humans have walked the Earth.

What holy book, what myth, can match the grandeur of that reality? In the face of such sublimity, why would any of us want to cling to the old tales, the comforting ones written thousands of years ago, the ones with all the answers but not many questions? No prophet ever imagined a universe built on such a scale, a cosmos so vast that fleeting light itself becomes a mere yardstick. Perhaps more wondrous than the enormity of the universe is the fact that we can actually measure its size and even pin down, to within a few tens of millions of years, its age.

As I write, two science stories, one exhilarating, one tragic, have managed to squeeze into headlines dominated by the threat of war and color-coded security warnings. One story concerns a NASA spacecraft — the Wilkinson Microwave Anisotropy Probe. From its orbit beyond the moon the probe has answered some of the biggest questions the human mind can pose. To an accuracy of 1 percent, we now know that the universe was born 13.7 billion years ago, and that it will probably expand forever. (On reflection, how could it have been otherwise? It's difficult to imagine something reversing and shutting down this whole explosive, radiant extravaganza.) The MAP spacecraft also confirmed that only about 4 percent of the matter in the universe is made from ingredients that we're familiar with — protons, neutrons, and electrons. Everything else in the cosmos consists of some unknown substance that physicists simply call dark matter, because they can't see it and have no clear idea about what it is. We are, it seems, made of exotic stuff.

The other story, the loss of the *Columbia* space shuttle, raises questions about the purpose of the space program and whether lives should be risked when we can design robotic spacecraft to explore the solar system and beyond. I have no doubt how *Colum-*

bia's astronauts would have answered that question. Some species evolved to remain anchored to one spot — barnacles, mussels, and molds come to mind — but not humans. Have you ever seen photos of the fossil hominid footprints from Laetoli, Tanzania? The prints were made 3.6 million years ago (a bit over the one-way lightspeed travel time to Andromeda) when two australopithecines walked through some volcanic ash. They always remind me of another collection of footprints, ones left in lunar dust just over thirty years ago. We've still got a lot of ground to cover, and the *Columbia*'s crew would not want us to stop now.

The hard-won new knowledge of the age and fate of the universe is something true and lasting. It doesn't depend on any particular culture's view of reality. It's not an opinion, or even a theory. It's the way things really are. The universe existed for a long time before we arrived to impose our visions of heaven and hell on it. Odds are that it will be around for a long time after we're gone. If you can stand another dose of cosmic humility, consider an illustration that appeared last January in the journal *Science*. The illustration showed the face of a clock, where each hour represented 1 billion years. Earth's first life appeared shortly before 1 A.M. Now we're at 4:30 A.M. In another half hour — 500 million years in real time — Earth's last living creatures, probably bacteria, will die, fried by a swollen sun. At noon the sun, by then a red giant, will have swallowed Mercury, Venus, and Earth. Not a comforting vision. But if confronting that reality challenges our beliefs and jolts our complacency, that is surely a good thing. The world has far too many people who, surfeited with murderous certainty, have stopped asking questions. How strange, though, that even those who scorn the worldview of science may nevertheless own computers, fly in passenger jets, use cell phones, or watch satellite television, apparently never considering that all these spring from the same source that tells us the universe is expanding and that life evolves.

It's too bad that the qualities necessary for good science — skepticism, secularism, and humanism — aren't as common as cell phones. Maybe someday they will be; we're still a young species. Meanwhile, those same qualities make for some unforgettable stories, no small solace in troubled times. In these pages you'll encounter Oliver Sacks asking if we are alone in the universe; Steven Weinberg, a Nobel laureate physicist, asks why our government is planning to spend billions of dollars on a missile defense plan that can't possibly work; Daniel Lazare shows that whatever else the Bible might be, it is not an accurate account of ancient history in the Near East; Natalie Angier, last year's guest editor, tells a surprising story about the vital evolutionary role of grandmothers; Elizabeth Loftus brings a psychologist's perspective to the sex scandals roiling the Catholic Church with an essay on the fallibility of memory. No doubt our new guest editor's introduction will further whet your appetite.

So, curious biped, sit, read, and wonder. Above all, wonder. That's what science is all about.

Even though most of a continent and all of an ocean separate us, working with Richard Dawkins has been a real treat for me. His books — *The Selfish Gene* and *The Blind Watchmaker* among them — are the rare sort that change the way you look at the world.

I would also like to remind readers of the passing last year of two remarkable men, David Wilkinson and Stephen Jay Gould. Wilkinson, a professor of physics at Princeton, was a key figure in the discovery that the universe began with the Big Bang. The MAP spacecraft, which might never have existed without Wilkinson's efforts, was renamed in his honor. When Gould, a paleontologist, gifted popularizer of science, and baseball fan, died in May 2002, the world lost an eloquent defender of evolution and rationalism.

Once again this year I'm in debt to Deanne Urmy and Melissa Grella at Houghton Mifflin for their help in pulling this book together. And for many years to come I hope to remain in the debt of Anne Nolan, the most supportive and lovely biped I have ever known.

TIM FOLGER

Introduction

IN INTRODUCING THIS ANTHOLOGY of American scientific writing I invoke two recently dead heroes, one a scientist and American, the other a writer, not trained in science and not from America but a lover of both. Carl Sagan gave one of his last books the characteristically memorable subtitle *Science as a Candle in the Dark*. Douglas Adams chose to study English literature at Cambridge, but he explained to me, in a televised conversation in 1997, that his reading habits have now changed: "I think I read much more science than novels. I think the role of the novel has changed a little bit. In the nineteenth century the novel was where you went to get your serious reflections and questionings about life. You'd go to Tolstoy and Dostoevsky. Nowadays, of course, you know the scientists actually tell us much, much more about such issues than you would ever get from novelists. So I think for the real solid red meat of what I read I go to science books, and read some novels as light relief."

Even while listening to him, I reflected on my frustration, going into bookshops and trying to find scientific books. If there is a science section at all, it is dwarfed not only by fiction, history, biography, "self-help," cookery, and gardening, but also by "new age," "occult," and religion. It has become a commonplace that astrology books outsell astronomy by a large margin.

Turning back to Adams, I asked him, "What is it about science that really gets your blood running?" and he replied: "The world is a thing of utter inordinate complexity and richness and strangeness that is absolutely awesome. I mean, the idea that such complexity can arise not only out of such simplicity but probably absolutely out of nothing is the most fabulous, extraordinary idea. And once you get some kind of inkling of how that might have happened — it's just wonderful. And I feel, you know, that the opportunity to spend seventy or eighty years of your life in such a universe is time well spent as far as I am concerned!"

Carl Sagan obviously shared those sentiments and devoted much of his career to expounding them, but *The Demon-Haunted World*, whose subtitle I quoted, has a darker theme. The darkness of ignorance breeds fear. In the words of a prayer which I early learned from my Cornish grandmother,

> From ghoulies and ghosties and long-leggety beasties
> And things that go bump in the night
> Good Lord deliver us.

Some say it is Scottish, not Cornish, but the sentiments are anyway worldwide. People are afraid of the dark. Science, as Sagan argued and personally exemplified, has the power to reduce ignorance and dispel fear. We should all read science and learn to think like scientists, not because science is useful (though it is), but because the light of knowledge is wonderful and banishes the debilitating and time-wasting fear of the dark. That uncompromisingly articulate chemist Peter Atkins has a utopian vision of a scientifically enlightened world which I share: "When we have dealt with the values of the fundamental constants by seeing that they are unavoidably so, and have dismissed them as irrelevant, we shall have arrived at complete understanding. Fundamental science can then rest. We are almost there. Complete knowledge is within our grasp. Comprehension is moving across the face of the Earth, like the sunrise."

Unfortunately, science arouses fears of its own, usually because of a confusion with technology. Even technology is not inherently frightening, but it can, of course, do bad things as well as good. If you want to do good, or if you want to do bad, science will provide the most effective way in either case. The trick is to choose the good rather than the bad, and what I fear is the judgment of those to whom society delegates that choice.

Science is the systematic method by which we apprehend what is true about the real world in which we live. If you want consolation, or an ethical guide to the good life, you can look elsewhere (and may be disappointed). But if you want to know what is true about reality, science is the only way. If there were a better way, science would embrace it.

Science can be seen as a sophisticated extension of the sense organs nature gave us. Properly used, the worldwide cooperative enterprise of science works like a telescope pointing toward reality; or, turned around, a microscope to dissect details and analyze causes. So understood, science is fundamentally a benign force, even though the technology that it spawns is powerful enough to be dangerous when abused. Ignorance of science can never be a good thing, and scientists have a paramount duty to explain their subject and make it as simple as possible (though no simpler, as Einstein rightly insisted).

Ignorance is usually a passive state, seldom deliberately sought or intrinsically blameworthy. Unfortunately, there do seem to be some people who positively prefer ignorance and resent being told the truth. Michael Shermer, debonair editor and proprietor of *Skeptic* magazine, tells of the audience reaction when he unmasked a professional charlatan onstage. Far from showing Shermer the gratitude he deserved for exposing a fake who was conning them, the audience was hostile. "One woman glared at me and told me it was 'inappropriate' to destroy these people's hopes during their time of grief."

Admittedly, this particular phony's claim was to communicate with the dead, so the bereaved may have had special reasons for resenting a scientific debunker. But Shermer's experience is typical of a more general mood of protective affection for ignorance. Far from being seen as a candle in the dark, or as a wonderful source of poetic inspiration, science is too often decried as poetry's spoilsport.

A more snobbish denigration of science can be found in some, but by no means all, literary circles. "Scientism" is as dirty a word as any in today's intellectual lexicon. Scientific explanations that have the virtue of simplicity are derided as "simplistic." Obscurity is often mistaken for profundity; simple clarity can be taken for arrogance. Analytical minds are denigrated as "reductionist" — as with "sin," we may not know what it means, but we do know that we are against it. The Nobel Prize–winning immunologist and polymath

Peter Medawar, not a man to suffer fools gladly, remarked that "reductive analysis is the most successful research stratagem ever devised," and continued: "Some resent the whole idea of elucidating any entity or state of affairs that would otherwise have continued to languish in a familiar and nonthreatening squalor of incomprehension."

Nonscientific ways of thinking — intuitive, sensitive, imaginative (as if science were *not* imaginative!) — are thought by some to have a built-in superiority over cold, austere, scientific "reason." Here's Medawar again, this time in his celebrated lecture "Science and Literature": "The official Romantic view is that Reason and the Imagination are antithetical, or at best that they provide alternative pathways leading to the truth, the pathway of Reason being long and winding and stopping short of the summit, so that while Reason is breathing heavily there is Imagination capering lightly up the hill."

Medawar goes on to point out that this view was even once supported by scientists themselves. Newton claimed to make no hypotheses, and scientists generally were supposed to employ "a calculus of discovery, a formulary of intellectual behaviour which could be relied upon to conduct the scientist towards the truth, and this new calculus was thought of almost as an antidote to the imagination."

Medawar's own view, inherited from his "personal guru" Karl Popper and shared by most scientists today, was that imagination is seminal to all science but is tempered by critical testing against the real world. Creative imagination and critical rigor are both to be found in this collection of contemporary American scientific literature.

For a non-American to be invited by a leading American publisher to anthologize American writings about science is an honor, the more so because American science is, by almost any index one could conjure, preeminent in the world. Whether we measure the money spent on research or count the numbers of active scientists working, of books and journal articles published, or of major prizes won, the United States leads the rest of the world by a convincing margin. My admiration for American science is so enthusiastic, so downright *grateful,* that I hope I may not be thought presumptuous if I sound a note of discordant warning. American science leads the world, but so does American *anti*-science. Nowhere is this more clearly seen than in my own field of evolution.

Evolution is one of the most securely established facts in all science. The knowledge that we are cousins to apes, kangaroos, and bacteria is beyond all educated doubt: as certain as our (once doubted) knowledge that the planets orbit the sun, and that South America was once joined to Africa, and India distant from Asia. Particularly secure is the fact that life's evolution began a matter of billions of years ago. And yet, if polls are to be believed, approximately 45 percent of the population of the United States firmly believes, to the contrary, an elementary falsehood: all species separately owe their existence to "intelligent design" less than ten thousand years ago. Worse, the nature of American democratic institutions is such that this perversely ignorant half of the population (which does *not,* I hasten to add, include leading churchmen or leading scholars in any discipline) is in many districts strongly placed to influence local educational policy. I have met biology teachers in various states who feel physically intimidated from teaching the central theorem of their subject. Even reputable publishers have felt sufficiently threatened to censor school textbooks of biology.

That 45 percent figure really is something of a national educational disgrace. You'd have to travel right past Europe to the theocratic societies around the Middle East before you hit a comparable level of antiscientific miseducation. It is bafflingly paradoxical that the United States is by far the world's leading scientific nation while *simultaneously* housing the most scientifically illiterate populace outside the Third World.

Sputnik, the Russian satellite launched in 1957, was widely seen as a salutary lesson, spurring the United States out of complacency and into redoubled educational efforts in science. Those efforts paid off spectacularly, for example, in the dazzling successes of the space program and the Human Genome Project. But more than forty years have passed since Sputnik, and I am not the only Americophile to suggest that another such fright may be needed. Short of that — well, in any case — we need excellent scientific writing for a general audience. Fortunately that high-quality commodity is in abundant supply in America, which has made the compiling of this anthology both easy and a pleasure. The only difficulty, indeed the only pain, has been in deciding what to leave out.

Should a collection such as this be timely or timeless? Topical and of-the-moment? Or *sub specie aeternitatis?* I think both. On the one hand, the volume is one of a series, tied to a particular year, sandwiched between predecessors and successors. That nudges us in the direction of topicality: what are the hot scientific subjects of 2003; what are the current political and social issues that scientific writings of the previous year might illuminate? On the other hand, science's ambitions — more so, I venture, than any other discipline's — approach the timeless, even the eternal. Laws of nature that changed from year to year, or even from eon to eon, would seem too parochial to deserve the name. Of course our *understanding* of natural law changes — for the better — from decade to decade, but that is another matter. And, within the unchanging laws of the universe, their physical manifestations change, on time scales spanning gigayears to femtoseconds.

Biology, like physics, anchors itself in uniformitarianism. Its defining engine — evolution — is change, change par excellence. But evolution is the same *kind* of change now as it was in the Cretaceous, and as it will be in all futures we can imagine. The play's the same, though the players that walk the stage are different. Their costumes are similar enough to connect, say, triceratops with rhinoceros, or allosaurus with tiger, in ecological continuity. If an ecologist, a physiologist, a biochemist, and a geneticist were to mount an expedition to the Cretaceous or the Carboniferous, their 2003-vintage skills and education would serve them almost as well as if they were going to, say, Madagascar today. DNA is DNA, proteins are proteins. They and their interactions change only trivially. The principles of Darwinian natural selection, of Mendelian and molecular genetics, of physiology and ecology, the laws of island biogeography, all these surely applied to dinosaurs, and before them to mammal-like reptiles, just as they apply now to birds and modern mammals. They will still apply in a hundred million years' time, when we are extinct and new faunistic players have taken the stage. The leg muscles of a tyrannosaur in hot-breathed pursuit were fueled by ATP such as any modern biochemist would recognize, charged up by Krebs cycles indistinguishable from the Krebs cycles of today. The science of life doesn't change from eon to eon, even if life itself does.

So far, so timeless. But we live in 2003. Our lives are measured in decades and our psychological horizons crammed somewhere between seconds and centuries, seldom reaching further. Science's laws and principles may be timeless, but science bears mightily upon our fleeting selves. The science and nature writing of 2002 is not the same as it was ten years ago, partly because we now know more about what is eternally true, but also because the world in which we live changes, and so does science's impact upon it. Some of the essays and articles in this book are firmly date-stamped; some are timeless. We need both.

Nature writing perennially returns to the theme of conservation and extinction. Of all arguments in favor of preserving species from extinction, I am moved more by aesthetic sentiment than by utilitarian advocacies of the "You never know whether something in the rain forest might eventually turn out to be useful to humanity" kind. But aesthetic isn't a big enough word, nor is sentiment. Douglas Adams's Professor Chronotis used his time machine for only one regular purpose: he would visit pre-seventeenth-century Mauritius, weep over the dodo, and return. The sense of irreparable loss — grief — our descendants will feel for elephants and whales brings today's imagination up short. Today we are still privileged to watch these great creatures, dodos for future generations to weep over. And we are still finding out new and extraordinary things about them, as "Four Ears to the Ground" and "Fat Heads Sink Ships" both show.

My personal dodo has long been the marsupial *Thylacinus,* often irritatingly called the Tasmanian tiger — irritatingly because it was much more like a dog (with a few stripes across the rump). I once wrote of it, "To any dog-lover, the contemplation of this alternative approach to the dog design, this evolutionary traveller along a parallel road separated by 100 million years, this part familiar yet part utterly alien other-worldly dog, is a moving experience. Maybe they were pests to humans, but humans were much bigger pests to them; now there are no thylacines left and a considerable surplus of humans." It is too late for the dodo, but "Raising the Dead" airs the faint hope (it may never reach the status of an expectation) of one day bringing *Thylacinus* back from the dead by cloning DNA from pickled museum specimens.

I once had the good fortune to spend two weeks in a tropical research center in Panama with my senior colleague and friend, the zoologist John Maynard Smith. We were being shown round by a young researcher whose enthusiasm moved Maynard Smith to whisper to me: "What a pleasure to listen to a man who really loves his animals." The "animals" in question were various species of palm tree. I was reminded of this when reading "Terminal Ice." It is all about men who love their animals, but their animals are icebergs. The article ends more grimly on what today's icebergs may be telling us about our globally warmed future. Complementing this article, "Ice Memory" tells how cores taken from glaciers constitute a sensitive record of climate changes of the past, perhaps foreshadowing an even grimmer future unconnected with global warming.

In my choice, I have been mindful that North America's natural heritage is perhaps the richest and most beautiful in the temperate world. It is also under threat from powerful interests more concerned with commercial exploitation than science, or beauty, or anything that we might recognize as civilized values at all. I do not, therefore apologize for including, among the natural history articles, some with a political agenda. These include "Maine's War on Coyotes" — and, by the way, on the subject of coyotophilia, I am sorry it was not possible to include extensive passages from Barbara Kingsolver's beautiful novel *Prodigal Summer.* "Sounding the Alarm" is a remembrance of the prophet Rachel Carson, and "The Bottleneck" a similar warning for our times from Edward O. Wilson.

Wilson is a scientific prophet if ever there was one, and I have also included a biographical piece representing him as a latter-day Thoreau, "Finding a Wild, Fearsome World Beneath Every Fallen Leaf." As another matched pair — article with biography of its author — I offer "The Fully Immersive Mind of Oliver Sacks" paired with Sacks himself on a slightly unexpected subject, "Anybody Out There?" The same theme, the possibility of extraterrestrial life, is treated rather differently by Tim Appenzeller in "At Home in the Heavens."

That title is a possibly unconscious allusion to Stuart Kauffman's otherwise very different *At Home in the Universe,* which in turn has weaker resonances with Timothy Ferris's *Coming of Age in the Milky Way.* Ferris himself is represented here by "Astronomy's New Stars," in praise of amateur astronomers. "The Very Best Telescope" pleases me because it presents a technical innovation as the solution to a problem: always my strategy when explaining the design of natural instruments such as eyes and echolocation systems. "A New

View of Our Universe" reaches the philosophical — some would say theological — cutting edge of cosmology. Is the universe not only our home but tailor-made for the task?

From theology sublime to theology mundane, "False Testament" reveals no great surprises, but the details are fascinating to those of us raised in a Judeo-Christian culture, and perhaps instructive, even salutary, to the benighted 45 percent that I mentioned earlier. Another archaeological piece is "Treasure Under Saddam's Feet." The dam that would flood these priceless antiquities to oblivion is due for completion in 2007. Might a halt to the damming plans turn out to be an unexpected benefit of war? I doubt it. In any case, war arouses greater fears for Iraq's other treasures, which rival those of Greece and Egypt in their archaeological importance.

How closely related are you to me? Probably closer than you think. My guess is based on the mathematics of Joseph Chang, discussed in "The Royal We." Most people have a natural curiosity about their ancestral past, and genetics is starting to develop methods to satisfy it, along with our sometimes morbid curiosity about our individual futures, as David Ewing Duncan discovers in "DNA as Destiny." Incidentally, those fearful that genetics may teach them too much about their own inexorable fates might take comfort from something we have known all along: identical twins don't habitually die on the same day. But how fated are we by our genes when it comes to abilities and talents? Steven Pinker, in "The Blank Slate," brings his customary acumen and style to dispelling the many misunderstandings that surround this question.

Pinker is identified with evolutionary psychology, one of those names — another being behavioral ecology — now used as a euphemism for what used to be called sociobiology. Natalie Angier's "Weighing the Grandma Factor" is a second piece in a genre that is regarded by some, for reasons that I understand but deny, as politically controversial. There's no denying, however, the controversy in some of the pieces I have chosen on scientific approaches to political or social questions. Steven Weinberg is one of the world's most distinguished physicists, and his "The Truth About Missile Defense" is an important document that should (but probably won't) be studied by politicians up to the highest level. Lawyers and judges should pay similar attention to Elizabeth Loftus's "Memory Faults and Fixes." Dr. Loftus is another scientific hero, whose courageous — and how sad that courage should be *necessary*— testimony on the sometimes inadvertent but more usually deliberate implanting of false memories has saved a significant number of innocent people from the current Salem-like hysteria over pedophilia.

"Embryo Police" is an American view of an institution that exerts considerable power in my own country, the Human Fertilization and Embryology Authority (HFEA). A famous case handled by the HFEA is that of Diane Blood, a young woman who tragically lost her husband to meningitis in 1995. While he was on a life support system, before his death, she persuaded the doctors to extract and freeze some of his sperm so that she might have his baby as they had always planned. The doctors obliged, but the HFEA subsequently denied her permission to undergo the in vitro fertilization on the grounds that her husband, in his terminal coma, could not give his written consent. After fighting them in the courts for years, Mrs. Blood was eventually allowed to take her husband's sperm abroad, and a European IVF clinic eventually gave her beloved husband two posthumous sons. She had to fight again to amend their birth certificates so their father was recorded as "Stephen Blood" rather than "Unknown." Perhaps unfairly, some might see Mrs. Blood's case as a cautionary tale from Britain for America, about the grief that can arise when lawyers and moralistic busybodies are given a license to poke their noses into private matters.

Diet is a political as well as a scientific issue, increasingly so as the epidemic of obesity gathers pace. Dr. Robert Atkins's long-running campaign to shift the blame from fats to carbohydrates is the subject of "What If It's All Been a Big Fat Lie?" I am not expert enough to give an authoritative verdict, but as a dispassionate observer I think it looks as though Atkins and his followers have built up a case that is at least compelling enough to demand a clear answer from that part of the medical establishment which once ridiculed him and now sounds desperate for his findings to go away.

The treatment of women in scientific careers was, until quite recently, often horribly unjust. The exclusion of Rosalind Franklin, she whose X-ray photographs were so crucial to Watson and Crick's discovery of the double helix, from the Common Room at King's College London, where her male colleagues could go and talk science, is infamous. I was reminded of it when I read "My Mother, the

Scientist," Charles Hirshberg's touching memoir of Joan Feynman, his mother (and the sister of the famous theoretical physicist). As Hirshberg puts it, "To become a scientist is hard enough. But to become one while running a gauntlet of lies, insults, mockeries, and disapproval — this was what my mother had to do. If such treatment is unthinkable (or, at least, unusual) today, it is largely because my mother and other female scientists of her generation proved equal to every obstacle thrown in their way."

In these more enlightened times, it is important to stop the pendulum overshooting the other way, such that young men find themselves at an unfair disadvantage in seeking employment in scientific or other academic work. Injustices to females in one era cannot be redressed today by injustices to males of a later era: it is a different lot of males and females! The same fallacy underlies ludicrous (and racist) demands for reparations to be paid to modern individuals with the same skin color as slaves by modern individuals with the same skin color as slaveowners.

Bill McKibben's "It's Easy Being Green" begins as a hymn of praise to his new car, a hyper-economical, environment-friendly hybrid-electric model. He modulates into baffled, and I think justified, anger against those in the motor industry and government who will not admit how easy and painless it would be to free ourselves from our "oil addiction" and "the gas-sucking SUVs that should by all rights come with their own little Saudi flags on the hood."

"Homeland Insecurity" is more technically interesting than angry, but it concerns an issue that will become ever more important to our society: how to protect information that all of us have a right to regard as private. This problem may become more controversial as we face up to government demands, in the face of the threat of terrorism, to encroach on our privacy.

The subject of terrorism brings me to my final choice, one for which I might need to offer an apology. The main talking point throughout 2002 was surely the appalling suicide attack on New York of the previous September, and it is hard to omit the topic from any collection of the year's writing. But what can science say to us at such a time? No doubt technology will offer ingenious suggestions for avoiding similar manmade tragedies in the future, from reinforced cockpit doors and automatic, pilot-free landing systems to stun grenades and narcotic drugs piped through airliner ventilation systems. But, true to my more academic view of science, I have chosen "A Skeptical Look at September 11th." It is only fair to mention that not everybody shares my enthusiasm for this piece by the space scientists Clark Chapman and Alan Harris. Some colleagues, whom I respect, are unconvinced by the assumptions underlying the authors' statistical estimates. Others even find the article offensive. But it seems to me that, far from downgrading the tragedy (which is the last thing anyone of goodwill wishes to do), Chapman and Harris offer constructive hope. If we all took a dose of their no-nonsense statistical common sense, bigots like bin Laden would be impotent (think of Germaine Greer's robust advice to women exposed to flashers: laugh at them). America is much too powerful for tinpot thugs like bin Laden to do widespread damage. But they can do enormous psychological harm if we let them unleash an epidemic of irrational fear.

Even if you disagree with Chapman and Harris's statistical reasoning, and even if you are offended by its timing, I would like to propose their essay as a type specimen, an example of how the scientific way of *thinking* might influence our lives for the better, quite apart from the more familiar ways in which scientific methods of *doing* can benefit us in practice. Franklin D. Roosevelt's famous remark "The only thing we have to fear is fear itself," which Chapman and Harris predictably quote, brings me back to my beginning: Carl Sagan's candle, and science's power to banish our fears of the dark.

RICHARD DAWKINS

The Best American Science
and Nature Writing 2003

NATALIE ANGIER

Weighing the Grandma Factor

FROM *The New York Times*

GRANDMA, WHAT A BIG and fickle metaphor you can be! For children, the name translates as "the magnificent one with presents in her suitcase who thinks I'm a genius if I put my shoes on the right feet, and who stuffs me with cookies the moment my parents' backs are turned."

In news reports, to call a woman "grandmotherly" is shorthand for "kindly, frail, harmless, keeper of the family antimacassars, and operationally past tense."

For anthropologists and ethnographers of yore, grandmothers were crones, an impediment to "real" research. The renowned ethnographer Charles William Merton Hart, who in the 1920s studied the Tiwi hunter-gatherers of Australia, described the elder females there as "a terrible nuisance" and "physically quite revolting" and in whose company he was distressed to find himself on occasion, yet whose activities did not merit recording or analyzing with anything like the attention he paid the men, the young women, even the children.

But for a growing number of evolutionary biologists and cultural anthropologists, grandmothers represent a key to understanding human prehistory, and the particulars of why we are as we are — slow to grow up and start breeding but remarkably fruitful once we get there, empathetic and generous as animals go, and family-focused to a degree hardly seen elsewhere in the primate order.

As a result, biologists, evolutionary anthropologists, sociologists, and demographers are starting to pay more attention to grandmothers: what they did in the past, whether and how they made a difference to their families' welfare, and what they are up to now in a sampling of cultures around the world.

At a recent international conference — the first devoted to grandmothers — researchers concluded with something approaching a consensus that grandmothers in particular, and elder female kin in general, have been an underrated source of power and sway in our evolutionary heritage. Grandmothers, they said, are in a distinctive evolutionary category. They are no longer reproductively active themselves, as older males may struggle to be, but they often have many hale years ahead of them; and as the existence of substantial proportions of older adults among even the most "primitive" cultures indicates, such durability is nothing new.

If, over the span of human evolution, postmenopausal women have not been using their stalwart bodies for bearing babies, they very likely have been directing their considerable energies elsewhere.

Say, over the river and through the woods. It turns out that there is a reason children are perpetually yearning for the flour-dusted, mythical figure called grandma or granny or oma or abuelita. As a number of participants at the conference demonstrated, the presence or absence of a grandmother often spelled the difference in traditional subsistence cultures between life or death for the grandchildren. In fact, having a grandmother around sometimes improved a child's prospects to a far greater extent than did the presence of a father.

Dr. Ruth Mace and Dr. Rebecca Sear of the department of anthropology at University College in London, for example, analyzed demographic information from rural Gambia that was collected from 1950 to 1974, when child mortality rates in the area were so high that even minor discrepancies in care could be all too readily tallied. The anthropologists found that for Gambian toddlers, weaned from the protective balm of breast milk but not yet possessing strength and immune vigor of their own, the presence of a grandmother cut their chances of dying in half.

"The surprising result to us was that if the father was alive or dead didn't matter," Dr. Mace said in a telephone interview. "If the grandmother dies, you notice it; if the father dies, you don't."

Importantly, this beneficent granny effect derived only from maternal grandmothers — the mother of one's mother. The paternal grandmothers made no difference to a child's outcome.

Dr. Donna Leonetti, an anthropologist at the University of Washington, and her colleague Dr. Dilip C. Nath presented similar results from their study of two contemporary ethnic groups in northeast India, one Bengali, the other Khasi. The two groups share certain fundamental characteristics, notably a heavy workload of manual labor, low income, and scant access to modern birth control methods.

But they differ in marital arrangements structure. Bengali wives move into their husbands' households, where they are supervised by their mothers-in-law. Khasi women, by contrast, stay in their natal homes, and their husbands join them.

The researchers discovered that for Bengali and Khasi families alike, having a grandmother around increased a young woman's overall fertility rate as compared with having no senior female on board. But the groups parted ways in the fate of the resulting offspring.

For the Bengali women, the paternal grandmother had no effect on the mortality rates of her grandchildren, with some 86 percent of children making it to age six whether the elder woman was there or not; while among the Khasi, 96 percent of children endured to age six if their maternal grandmothers were alive, compared with only 83 percent if the grandmother had died.

The researchers cannot explain what, exactly, these grand old doyennes are doing. One presumed measure of viability, a child's growth rate, does not differ significantly between Khasi children with living grandmothers and those without.

Indeed, a number of researchers at the conference admitted to being flummoxed by the nature of grandma's goodness.

"This was a constant refrain: what is the mechanism?" said Dr. Patricia C. Draper, a professor of anthropology at the University of Nebraska. "We can see that grandmothers are doing something, but what? What buttons are they pushing that end up making the difference to their families?"

Perhaps, she suggested, they exerted as much of a psychological as a practical effect — for example, by encouraging family cohesion or stifling extreme sibling rivalry.

In a couple of studies, the divergent effects of the two grandmother species is so pronounced that the son's mother appears not merely a neutral influence on her grandchildren, but a negative one.

Dr. Cheryl Jamison, an anthropologist at Indiana University in Bloomington, and her colleagues combed through an exceptionally complete population register from a village in central Japan. The records covered a period from 1671 to 1871, when officials sought to battle the encroachment of Christianity and thus kept track of everybody's birth, death, and whereabouts, the better to interrogate citizens each year on their religious allegiance. As in the Gambian study, the overall mortality rate for children was substantial, with 27.5 percent of children dying by age sixteen.

Dr. Jamison and her coworkers determined that when a maternal grandmother lived in the household, boys were 52 percent less likely to die in childhood than if there was no grandmother present. Conversely, when the father's mother lived in the house, boys were 62 percent more likely to die than were those without a resident grandma. For girls, no statistically significant benefit or decrement could be seen from grandmothers of either bloodline.

Dr. Jamison cautioned that not too much could be made of the results, for, in a patrilineal culture like that of premodern Japan, where sons were the ones who took in their aging parents, the sample size for maternal grandmothers living with their grandchildren was extremely small compared with that of coresiding paternal grandparents and young children. Nevertheless, she said, she was startled by her results.

"One would think that boys would be preferred by everybody, but apparently that isn't what happened here," she said in an interview.

Why boys should be helped or harmed to a comparatively greater extent than are female children in this sample could not be gleaned from the population registry data, Dr. Jamison said.

Those researchers with a Darwinian bent propose that the discrepant effects of maternal versus paternal grandparents are a result of the old evolutionary bugaboo, paternity uncertainty. Maternal grandmothers, they reason, are confident that the grandchildren in question are their blood relations, and hence worth working for, whereas the mother of a son, ever unsure of her daughter-in-law's fidelity, may withhold her love and care, albeit unconsciously, from the young bairns before her.

Dr. Harald A. Euler, a professor of psychology at the University of Kassel in Germany, believes that the maternal-paternal grandparent divide continues to this day and affects our affections.

He interviewed 2,000 people in Germany about their grandparents — how much care they received from the relatives, how
much affection they had for each one — and then analyzed the responses of the 700 who had all four grandparents alive until age
seven or later. He found, among other results, that half the respondents cited their maternal grandmother as their favorite grandparent, while only 12 to 14 percent named the paternal grandmother.

His research, he said, has helped him come to terms with what
he once thought was an inexcusable indifference on his mother's
part toward his two children, especially compared with the interest
displayed by his wife's mother.

"I thought, what a coldhearted woman," he said. "Now, I don't
blame her anymore."

Others are less quick to pin everything on biology, and offer another explanation for the comparatively salubrious effects of a maternal grandmother. Mothers overwhelmingly are the designated
caretakers of children, they say, and when children need help,
whose name are they going to call? Mom-my!

"It's to be expected that a woman would turn to the person
she knows best for help with the children, and that person is much
likelier to be her mother than her mother-in-law," said Dr. Martin Kohli, director of the Research Group on Aging and the Life
Course at the Free University of Berlin. "And so it is that the maternal lineage has the opportunity to make a difference."

Dr. Kohli said a new French study of contemporary grandparent-
hood had found, among other things, that paternal grandparents
often wanted to do more for their grandchildren, but felt they were
not as welcome to visit as were the maternal grandparents.

Indeed, Dr. Kohli's own research has revealed the degree to
which grandparents on each side of the family are keen to make a
difference in the lives of their descendants. Analyzing financial
data in Germany and comparing it with similar investigations in
the United States and France, Dr. Kohli concluded that contrary to
the image of older people as a financial drain on their younger relatives, the net flow of money was practically one-directional, from
the old to the young.

In 1996, for example, Germans over the age of sixty-five gave
$3,600 more to their junior relations than they received in gifts of
money or other goods. Even those seniors living on tight fixed in-
comes scrimped enough to give to their kin, setting aside an aver-

age of 9 percent of their government pensions to donate to their children and grandchildren, with no measurable distinction between whether they were the children of their sons or of their daughters.

"One thing is for sure, we're not talking greedy geezers here," Dr. Kohli said. "The majority are quite anxious to give something, to leave something, to help their descendants."

Who better to keep the wolf from the door than Grandma?

TIM APPENZELLER

At Home in the Heavens

FROM *U.S. News & World Report*

WAIMEA, HAWAII — Just after midnight, Geoff Marcy and Paul Butler turn on the New World Symphony and take the controls at the Keck Telescope headquarters, here on Hawaii's Big Island. The music suits their plan for the night: spending the wee hours searching for planets around other stars. What they and scores of other astronomers most want to find, however, is not a strange new world but one that looks a little bit like home.

After fifteen years on the quest, Marcy and Butler have it down to, well, a science. That night in late February, their program ran to fifty stars. Over a video link from the control room to an operator at the Keck I telescope itself, 48 miles away on 13,800-foot Mauna Kea, they took aim at star after star. From each one the 10-meter-wide telescope — the world's largest along with its twin, the Keck II — captured a few minutes of light, to be analyzed for hints of unseen giant planets tugging the star back and forth.

This "wobble" technique, though indirect, has been hugely fruitful since a Swiss group found the first alien planet in 1995. By early this year, planet hunters had discovered about eighty. But the gargantuan size of the planets — up to ten times the mass of Jupiter and more — and their freakish orbits, often smaller than Mercury's or drastically elongated, have made them utterly unlike the nine planets we know. "The planets we're finding are fun and wacky," Butler, of the Carnegie Institution of Washington, said in February. "But nothing looks like our solar system."

Wacky planets fascinate astronomers, but the question they most want to answer is: Could there be other havens for life? Since it will be years before they can hope to detect Earth-size planets, that means looking for solar systems resembling our own, where small, stable planets might survive. Some observers are pushing telescope technology to the limits to see alien giant planets directly. Others are building on the tried-and-true wobble technique. That night in Hawaii, Marcy and Butler made some of the final observations that paid off late last week when their team reported fifteen more planets. And one of them, around a star called 55 Cancri, has a familiar look.

"It's the first analog of a solar system giant planet," says Marcy, who is at the University of California–Berkeley. Although the new planet weighs at least three and a half times as much as Jupiter, it follows a similar orbit: nearly circular and about as far from its sun. And newer search strategies promise more direct hints of home, by capturing the shadows of giant planets crossing the face of their suns or the glow of hot, newly formed Jupiters. If alien planets can actually be imaged, says Berkeley astronomer Ray Jayawardhana, "they will have become real worlds."

Big prospects. Giant planets are unlikely havens for life. But finding them in Jupiter-like orbits brightens the prospect that planets like our own might be circling the same stars. The giants' wide, circular paths would keep them far from vulnerable small planets closer in, and their powerful gravity would sweep up asteroids and comets that might pummel a terrestrial planet and snuff out any life.

It took years of observing before Marcy, Butler, and their colleagues were ready to announce their Jupiter, because their technique depends on watching a star get pulled to and fro through a planet's full orbit. A Jupiter takes twelve years or more. Yet this one, forty-one light-years away, is still an imperfect match to our own. The same star has two other giant planets, one of them a Jupiter-size body in an orbit so small that it completes a revolution every 14.6 days. It probably got there by migrating inward from the planetary system's outer reaches soon after it formed. That could mean no Earths, says Carnegie theorist Alan Boss. "Bodies that were going to form terrestrial planets undoubtedly got kicked out of the way by this big guy that came muscling through." But other Jupiters, without heavyweight companions, may be lurking among the hundreds of other stars that Marcy, Butler, and others have been watching. A couple more years of observations may tease them out.

Even before then, other astronomers, including Jayawardhana, may short-circuit the search by taking a picture of an alien Jupiter. A giant planet near a young star — less than 100 million years old, a baby compared with our 4.6-billion-year-old sun — would still be hot from its formation, glowing with infrared light. It could be just bright enough to register in a giant telescope like the Keck, equipped with "adaptive optics" technology that detects and cancels out the atmosphere's blurring. The discovery would signal a solar system in the making, and light from the planet might show the fingerprints of the same gases seen in our own Jupiter, including water, methane, and ammonia.

On the same February night while Marcy and Butler were at the controls of Keck I, Jayawardhana, David Charbonneau of the California Institute of Technology, and Brigitte Koenig, a Ph.D. student at the Max Planck Institute in Germany, were in the adjoining control room, observing on Keck II. They were working through a list of young, sunlike stars within a hundred light-years or so of Earth. One star on the agenda was especially tantalizing, because earlier observations on the Hubble Space Telescope had revealed a faint point of light right next door. It could be a planet. But it could also be a star much farther away that happened to be lined up just right to fool astronomers. One way to tell is to take a second look to see if the spot wanders across the sky in tandem with its star, as a planet would be expected to do.

One by one, the planet hunters inspected their stars. Each time the telescope locked on to a new target, there was a pause while the adaptive optics system got to work. It bounced the starlight off a flexible mirror that bends, twists, and quivers hundreds of times a second to cancel out atmospheric turbulence. After each star flared into view on a monitor, the astronomers tapped commands to maneuver a tiny light shield inside the detector until it masked the star, making it easier to see any faint neighbor.

Nothing planetlike turned up that night. Weeks later, after Jayawardhana and his colleagues had processed the data to sharpen the images, they concluded that the promising spot was a background star. But last week they were back at the Keck, scanning more young stars for the glimmer of a planet. "We're pushing the limits of what's technically possible, so we know the prize won't come easy," says Jayawardhana. "One of us will find the first planet sometime soon, and that will be fantastic."

The shadow. In fact, other astronomers have actually seen a planet around another star — though not as an image but as a shadow. Marcy, Butler, and Steve Vogt of UC–Santa Cruz had detected the planet indirectly, from the wobbles of its parent star. Then other astronomers saw the star dimming like clockwork every three and a half days as the close-orbiting planet swung past it.

When Charbonneau and others analyzed the dimmed starlight with the Hubble telescope last year, they saw something else: a subtle color change. The starlight was shining through the planet's atmosphere, picking up the tint of sodium vapor, a trace constituent. It was astronomers' first glimpse of what an alien world is made of.

In the next year, Charbonneau and his colleagues plan to study this planet's atmosphere with the Hubble and the Keck, looking for hints of water vapor, methane, and other gases. He and others will also be searching for more planetary shadows with small, wide-angle telescopes, able to watch tens of thousands of stars at once. Already, astronomers have seen a handful of dimmings that might be planets — though it will take follow-up observations on bigger telescopes to be sure.

If they are planets, the shadows should hold clues to their size and makeup. And worlds unimaginably far away will seem a little more comprehensible, and perhaps a little more like our own.

ALAN BURDICK

Four Ears to the Ground

FROM *Natural History*

FROM TIME TO TIME, leaving the American Museum of Natural History after hours, I pass the elephants in the Akeley Hall of African Mammals. They occupy the center of the room: a cluster of them, on a wide dais, milling eternally in the state of taxidermy. Aside from them and me and a savanna of glass-eyed ungulates, the hall is empty. My footsteps produce the only sound, which seems somehow amplified by the elephants' great mass.

We share a regular, wordless dialogue, the elephants and I, but only lately have I come to understand what they have to say. For years now, scientists have understood that elephants communicate at a frequency typically too low for the human ear to perceive — about twenty hertz. Propagating through the air, these vocal calls can reach an elephant five miles away. For better reception, the listening elephant spreads its earflaps forward, effectively transforming its head into a satellite dish.

As it turns out, that is only half the story. Recently a Stanford University researcher, Caitlin O'Connell-Rodwell, discovered that an elephant's vocal call actually generates two separate sounds: the airborne one and another that travels through the ground as a seismic wave. Moreover, the seismic version travels at least twice as far, and seismic waves generated by an elephant stomping its feet in alarm travel farther still, up to twenty miles. What's most remarkable, however, is how elephants presumably perceive these signals: they listen, it seems, with their feet.

Seismic communication is widespread. Creatures from scorpions to crocodiles rely on ground vibrations to locate potential mates and to detect (and avoid becoming) prey. The male fiddler crab bangs territorial warnings into the sand with its oversized claw. A blind mole rat pounds its head against the walls of its underground tunnels, thus declaring its dominance over the blind mole rat two tunnels over, which may or may not be listening with its own head pressed to the wall.

O'Connell-Rodwell was first inspired by the seismic songs of planthoppers, tiny insects she studied early in her career. The planthopper sings by vibrating its abdomen; this causes the underlying leaf, and ideally all nearby planthoppers, to tremble. She observed that planthoppers in the peanut gallery would lift a foot or two, presumably for better hearing: the other feet, bearing more weight, thus became more sensitive to vibration. Years later, O'Connell-Rodwell saw similar behavior among elephants at a water hole in Namibia. Minutes before a second herd of elephants arrived, members of the first group would lean forward on their toes and raise a hind leg, as if in anticipation. "It was the same thing the planthoppers were doing," she says.

Was it? Several elegant experiments by O'Connell-Rodwell demonstrate that elephants do indeed generate long-range seismic signals. But can other elephants hear them? Early evidence from northern California's Oakland Zoo, where an elephant named Donna is being trained to respond exclusively to seismic cues, strongly suggests that the answer is yes. "We haven't sealed the deal," says O'Connell-Rodwell, "but it looks promising."

As a communication medium, she notes, seismic waves would offer the elephant several advantages. They dissipate less quickly than airborne waves, they aren't disrupted by changes in weather or temperature, and they aren't swallowed by dense jungle foliage. Complex vocal harmonics don't translate well into seismic waves. But even the simplest long-range message — "I'm here" or "Danger!" — beats a fancy one that can't be heard at all.

Air is the faster medium: an airborne elephant call will reach a distant listener before the seismic version does. The delay between signals may confer its own advantage, however, O'Connell-Rodwell proposes. The delay increases with distance; an astute listener would soon learn to gauge distance from the delay. Combined with its airborne counterpart, a seismic signal would enable the animal to coordinate its movements with faraway colleagues, to forage more effectively, and to detect unseen danger. It is compass, yard-stick, and e-mail in one — an elephantine Palm Pilot.

And the elephant's palm is the key, O'Connell-Rodwell believes. It may be that the seismic vibrations propagate from the elephant's feet to its inner ear — a process known as bone conduction. That would explain some of the odder features of elephant anatomy, including the fatty deposits in its cheeks, which may serve to amplify incoming vibrations. In marine mammals, similar deposits are called "acoustic fat."

But O'Connell-Rodwell thinks the elephant ear may be tuned even more acutely to the ground. "They do have nerves connected to their toenails, and they do lean on them. It could be a direct line to their head." A colleague is now exploring whether the fleshy pad of an elephant's foot contains Pacinian and Meissner corpuscles, specialized nerve endings that detect faint motion and vibration. The tip of an elephant's trunk has more of these structures per square inch than does any other animal organ, and it is supremely touch-sensitive. (In addition to lifting a foot to improve its hearing, an elephant sometimes holds its trunk to the ground, as if it were an amplifier, says the Stanford biologist.)

All of which raises the question, Which is doing the hearing here — the elephant foot or the elephant ear? The truth is, "hearing" is a semantic distinction, a construct of human language. To us, a "sound" is what happens when airborne acoustic waves vibrate tiny hairs inside our head. An "ear" is an acoustic organ that looks like ours.

Properly defined, however, sound is a series of compression waves in any medium: air, liquid, solid matter. Animals have evolved all manner of translating these mechanical waves into neural signals. A fish senses motion with a line of specialized receptors on both sides of its body. Walk toward a fish tank, and your footsteps startle the fish. Did it hear you or feel you? To the fish, there's no difference.

Perhaps, in our ear-o-centric view of the world, we have constrained our senses. "The animals have been paying attention to something that we haven't been noticing," O'Connell-Rodwell says. Lately she has begun exploring the possibility that other large mammals — bison, rhinoceroses, hippopotamuses, lions, giraffes — rely on seismic cues in their daily lives.

Paradoxically, the discovery that elephants and perhaps other large mammals may communicate seismically comes at a time when it is increasingly difficult for us to hear them. Just as the night sky is slowly becoming obscured by "light pollution" from countless streetlights and other artificial sources of illumination, so the sounding board of earth has become muddled with "bioseismic noise": rumbling trucks, electric generators, jet vibrations, the hum and trundle of civilization and commerce. Does this human static disrupt elephant conversations in the wild? Does it drive them nuts in captivity? The zoo environment is stressful enough without having to hear from every pothole within a twenty-mile radius. Then again, I manage to sleep through the most fearsome Manhattan traffic. "My guess is, elephants in urban environments have become desensitized to seismic signals, as people have," suggests O'Connell-Rodwell.

In the end, the primary casualty of bioseismic noise is us. The human foot happens to be a remarkably sensitive listening device. It is nearly as dense with pressure receptors as is the elephant's trunk. O'Connell-Rodwell suspects that once upon a quieter time, we paid closer attention to seismic signals than we do today. Vibrations from instruments such as the talking drum or the didgeridoo, or even from foot-stomping dances, may have spoken volumes to distant, unshod listeners. Then came telephones, automobiles, asphalt — and footwear. We hardened our soles to the world of sound.

The echo of my footsteps haunts me now. When last I strolled through the darkened Akeley Hall, it struck me that this is what it would be like to be entombed in a shoe. The silent elephants, the hushed lions, the stilled giraffes — a continent of primordial instincts urged me toward the exit: Loosen, unlace, enter the world barefoot.

CLARK R. CHAPMAN
AND
ALAN W. HARRIS

A Skeptical Look at September 11th

How We Can Defeat Terrorism by
Reacting to It More Rationally

FROM *Skeptical Inquirer*

HUMAN BEINGS might be expected to value each life, and each death, equally. We each face numerous hazards — war, disease, homicide, accidents, natural disasters — before succumbing to "natural" death. Some premature deaths shock us far more than others. Contrasting with the 2,800 fatalities in the World Trade Center (WTC) on September 11, 2001 (9/11), we barely remember the 20,000 Indian earthquake victims earlier in 2001. Here, we argue that the disproportionate reaction to 9/11 was as damaging as the direct destruction of lives and property. Americans can mitigate future terrorism by learning to respond more objectively to future malicious acts. We do not question the visceral fears and responsible precautions taken during the hours and days following 9/11, when there might have been even worse attacks. But, as the first anniversary of 9/11 approaches, our nation's priorities remain radically torqued toward homeland defense and fighting terrorism at the expense of objectively greater societal needs. As we obsessively and excessively beef up internal security and try to dismantle terrorist groups worldwide, Americans actually feed the terrorists' purposes.

Every month, including September 2001, the U.S. highway death toll exceeds fatalities in the WTC, Pentagon, and four downed airliners combined. Just like the New York City firefighters and restaurant workers, last September's auto crash victims each had families, friends, critical job responsibilities, and valued positions in their churches and communities. Their surviving children, also, were left without one parent, with shattered lives, and much poorer than the 9/11 victims' families, who were showered with $1.5 million, per fatality, from the federal government alone. The 9/11 victims died from malicious terrorism, arguably compounded by poor intelligence, sloppy airport security, and other failed procedures we imagined were protecting us. While few of September's auto deaths resulted from malice, neither were they "natural" deaths: most also resulted from individual, corporate, and societal choices about road safety engineering, enforcement of driving-while-drunk laws, safe car design, and so on.

A Lack of Balance

Why does 9/11 remain our focus rather than the equally vast carnage on the nation's highways or Indian earthquake victims? Some say, "Oh, it was a natural disaster and nothing could be done, while 9/11 was a malicious attack." Yet better housing in India could have saved thousands. As for malice, where is our concern for the 15,000 Americans who die annually by homicide? Apparently, the death toll doesn't matter, not if people die all at once, not even if they die by malicious intent. We focus on 9/11, of course, because these attacks were terroristic and were indelibly imprinted on our consciousness by round-the-clock news coverage. Our apprehension was then amplified when just a half-dozen people died by anthrax. Citizens apparently support the nation's sudden, massive shift in priorities since 9/11. Here, we ask "Why?"

Suppose we had reacted to 9/11 as we did to last September's auto deaths. That wouldn't have lessened the destroyed property, lost lives and livelihoods, and personal bereavement of family and associates of the WTC victims. But no billions would have been needed to prop up airlines. Local charities wouldn't have suffered as donations were redirected to New York City. Congress might have enacted prescription drug benefits, as it was poised to do before 9/11. Battalions of National Guardsmen needn't have left their jobs to provide a visible "presence" in airports. The nation might not have slipped into recession, with resulting losses to businesses, workers, and consumers alike. And the FBI might still be focusing on rampant white-collar crime (think Enron) rather than on terrorism. While some modest measures (e.g., strengthening cockpit doors) were easy to implement, may have inhibited some "copycat" crimes, and may even lessen future terrorism, we believe that much of the expensive effort is ineffective, too costly to sustain, or wholly irrelevant.

Some leaders got it right when they implored Americans after 9/11 to return to their daily routines, for otherwise "the terrorists will win." Unfortunately, such exhortations seemed aimed at rescuing the travel industry rather than articulating a broad vision of how to respond to terrorism. We advocate that most of us more fully "return to normal life." We suggest that the economic and emotional damage unleashed by 9/11, which touched the lives of all Americans, resulted mostly from our *own* reactions to 9/11 and the anthrax scare, rather than from the objective damage. We recognize that our assertion may seem inappropriate to some readers, and we are under no illusion that natural human reactions to the televised terrorism could have been wholly averted and redirected. We, too, gaped in horror at images of crashing airplanes, and we contributed to WTC victims. But from within the skeptical community there could emerge a more objective, rational alternative to post-9/11. Citizens could learn to react more constructively to future terrorism and to balance the terrorist threat against other national priorities. It could be as important to combat our emotional vulnerability to terrorism as to attack Al Qaeda.

Terrorism, by design, evokes disproportionate responses to antisocial acts by a malicious few. By minimizing our negative reactions, we might contribute to undermining terrorists' goals as effectively as by waging war on them or by mounting homeland defenses. We do *not* "blame the victims" for the terrorists' actions. Rather, we seek that we citizens, the future targets of terrorism, be empowered. As Franklin D. Roosevelt famously said, "The only thing we have to fear is fear itself." We can help ensure that terrorists don't win if we can minimize our fears and react more constructively to future terrorism. We don't suggest that this option is easy or will suffice alone. It may not even be possible. But human beings often best succeed by being rational when their emotions, however tenacious and innate, have let them down.

Death and Statistics

It is a maxim that one needless or untimely death is one too many. So 20,000 victims should be 20,000 times worse. But our minds don't work that way. Given the national outpouring of grief triggered by the estimated 6,500 WTC deaths, one might have expected celebration in late October when it was realized that fewer than half that many had died. But there were no headlines like "3,000 WTC Victims Are Alive After All!" The good news was virtually ignored. Weeks later, many — including Defense Secretary Donald Rumsfeld — continued to speak of "over 5,000 deaths" on 9/11.

To researchers in risk perception, this is natural human behavior. We are evolved from primitive nomads and cave dwellers who never knew, personally, more than the few hundred people in their locales. Until just a few generations ago, news from other lands arrived sporadically via sailors; most people lived and died within a few miles of where they were born. Tragedies invariably concerned a known, nearby person. With the globalization of communication, the world — not just our local valley — has entered our consciousness. But our brains haven't evolved to relate, personally, to each of 6 billion people. Only when the media single out someone — perhaps an "average layperson" or maybe a tragic exception like JonBenet Ramsey — do our hearts and minds connect.

When an airliner crashes, and reporters focus on a despairing victim's spouse or on the last cellular phone words of a doomed traveler, our brains don't think statistically. We imagine ourselves in that airplane seat, or driving to the airport counseling center when our loved one's plane is reported missing. Actually, 30,000 U.S. commercial flights occur each day. In 2001, except for September 11 and November 12 (when an airliner crashed in Queens, New York, killing more passengers and crew than the four 9/11 crashes combined), no scheduled, U.S. commercial air trips resulted in a single passenger fatality. Indeed, worldwide airline accidents in 2001 — including 9/11 — killed fewer passengers than during an average year. But statistics can't compete with images of emergency workers combing a crash site for body parts with red lights flashing. We are gripped by fear as though the tragedy happened in our own neighborhood, and another might soon happen again.

Some responses to 9/11 were rational. Soon after jumbo jets were used as flying bombs, workers in landmark skyscrapers might reasonably have feared that their building could be next. With radical Muslims preaching that Americans must be killed, it might behoove us to avoid conspicuous or symbolic gatherings like Times Square on New Year's Eve or the Super Bowl. Surely disaster managers must plug security loopholes that could permit thousands or millions more to be killed. But when police chiefs of countless Middle American communities beef up security for *their* anonymous buildings and search fans entering hundreds of sports fields to watch games of little note, official reactions to terrorism have run amok. To imagine that Al Qaeda's next target might be the stadium in, say, Ames, Iowa, is far-fetched indeed.

Finite Resources, Infinite Alarm

Americans' WTC fears only grew when six people died from mailed anthrax. Postal officials patiently explained that public risks were minimal. But millions donned gloves to open their mail or gingerly threw out unopened mail; post offices rejected letters lacking return addresses; urgent mail was embargoed; and for weeks the national dialogue centered on one of the least hazards we face. An NPR radio host asked the postmaster general if the whole U.S. Postal System might be shut down, despite expert opinion that — in a world faced with diabetes, salmonella poisoning, and AIDS — anthrax will remain (even as a biological weapon) a bit player as a cause of death. Its sole potency is in the context of terrorism: if, by mailing lethal powder to someone, the news media choose to broadcast hysteria into every home so that the very future of our postal system is questioned, then the terrorist has deployed a powerful weapon indeed. But his power would be negated if we were to react to the anthrax in proportion to its modest potential for harm.

Research on risk perception has shown that our reactions to hazards don't match the numerical odds. We fear events (like airliner crashes) that kill many at once much more than those that kill one at a time (car accidents). We fear being harmed unknowingly (by carcinogens) far more than by things we feel we control ourselves (driving or smoking). We fear unfamiliar technologies (nuclear power) and terrorism far more than prosaic hazards (household falls). Such disproportionate attitudes shape our actions as public citizens. Accordingly, governments spend vastly more per life saved to mitigate highly feared hazards (e.g., aircraft safety) than on "everyday" risks (e.g., food poisoning). Risk analysts commonly accept, with neutral objectivity, the disparity between lay perceptions and expert risk statistics. Sometimes it is justifiable to go beyond raw statistics. Depending on our values, we might be more concerned about unfair deaths beyond an individual's control than self-inflicted harm. We might worry more about deaths of children than of elderly people with limited life expectancies. We might dread lingering, painful deaths more than sudden ones. We might be more troubled about "needless" deaths, with no compensating offsets, than about fatalities in the name of a larger good (e.g., soldiers or police). Or, in all these cases, we might not.

Why should terrorism command our exceptional attention? That the 9/11 terrorists maliciously attacked the symbolic *and actual* seats of our economic and military power (WTC/Wall Street and the Pentagon) *should* concern us if we truly think that future attacks might destroy our society. But who believes that? Government responses seem directed mostly at stopping future similar attacks . . . which returns us full circle to the question: why should that have become our primary national goal, at the sacrifice of tens of billions of dollars, of some of our civil liberties, of our traveling convenience, and of many of our pre-9/11 priorities?

Instead of rationally apportioning funds to the worst or most unfair societal predicaments, homeland security budgets soar. Nearly every airport administrator, city emergency management director, mayor, legislator, school district supervisor, tourist attraction manager, and plant operations foreman felt compelled after 9/11 to "cover their asses" by visibly enhancing their facility's security. Superfluous barricades were erected, search equipment purchased, and guards hired. Postage rates and delivery delays increased as envelopes were searched for anthrax. Even the governor of West Virginia announced a "West Virginia Watch" program; while some vigilance in that state does no harm, it is unlikely that Wheeling is high on Osama bin Laden's target list.

Meanwhile, programs unrelated to "homeland security" suffered. Finite medical resources were diverted to comforting people that their flu symptoms weren't anthrax . . . or testing to see if they were. Charitable funds that would have nurtured the homeless flowed, instead, to wealthy families of deceased Wall Street traders. Funds for education and pollution control go instead to "securing" public buildings and events. Billions of extra tax dollars are spent on military operations in Pakistan and Afghanistan rather than on enhancing American productivity. If we truly believe in "life, liberty, and the pursuit of happiness" and that each life is precious, we must resist selfish forces that would take advantage of our fears and squander our energies and fiscal resources on overblown security enhancements.

Many say that spending for extra security can do no harm. But there *is* harm when politicians act on views, like those of a New Yorker who earlier this year disparaged complaints about airport queues, saying, "I hope that they will be inconvenienced, and will always be inconvenienced, because we should never forget the 5,000 [*sic*] who died." "Inconvenience" sounds innocuous, but it means lost time, lost money, lost productivity, as well as increased frustration and cynicism. Disproportionate expenditures on marginal security efforts take attention, time, and resources away from other more productive enterprises.

Moreover, our civil liberties are eroded by the *involuntary* nature of our "sacrifices." When a person irrationally fears crowded elevators and takes the stairs instead, only that person suffers the inconvenience of their personal response. But when everyone, fearful or not, is forced to suffer because of the fears of others, then such measures become tyrannical: we should expect rational deliberation and justifications by our leaders before accepting them. But in the aftermath of 9/11, tens of billions of dollars were immediately reallocated with little public debate. Skeptics might well question our society's acquiescence to popular hysteria and proactively challenge our leaders to balance the expenditures of our resources.

Misperceptions of Risk

Consider some misperceptions of risk. Many news headlines just before 9/11 concerned shark attacks and the disappearance of Chandra Levy, an extreme distortion of serious societal issues (only ten people annually are killed by sharks worldwide). We can laugh at, or bemoan, the triviality of the media. But such stories reflect our own illogical concerns. If, in allocating funds among different hazards, we deliberately *choose* to value the lives of Manhattan sky-scraper office workers, postal employees, or airline frequent flyers more than we value the lives of agricultural workers or miners, it is a conscious, informed choice. But it is rarely objectivity that informs such choices. In order to help laypeople and leaders to put our options into perspective, skeptics, teachers, and journalists alike have a responsibility to put the objective past and potential threats from terrorism into contexts that ordinary people can relate to.

Let's compare 9/11 with other past and potential causes of mass death. Note that we generally can't compare prevention costs with lives saved; at best, we can compare expenditures with lives *not* saved. For example, we can compare the cost of air traffic control with midair collision fatalities, but we can only guess at the toll without any such air traffic control.

- 9/11 deaths are similar to monthly U.S. traffic fatalities. Whatever total private/public funds are spent annually, per life saved, on improved highway and motor vehicle safety, alcohol-while-driving prevention efforts, etc., it hardly approaches homeland security budgets.
- The 9/11 fatalities were several to ten times fewer than annual deaths from falls (in the home or workplace), or from suicide, or from homicide. One can question the effectiveness of specific safety programs, counseling efforts, or laws; but, clearly, comparatively paltry sums are spent on programs that would further reduce falls, suicides, and murders.
- In autumn 2001, the Centers for Disease Control and Prevention (CDC) predicted that 20,000 Americans would die from complications of influenza during the then-upcoming winter, most of which could be prevented if susceptible people were vaccinated. The CDC advisory was typically buried inside newspapers whose banner headlines dealt with the anthrax attacks, which killed just a few people.
- Twice as many people died in the worst U.S. flood (stemming from the 1900 Galveston hurricane) as at the WTC. Floods and earthquakes are major killers abroad (each of ten disasters killed over 10,000 people, and a few over 100,000, during the last three decades, chiefly in Asia) but are minor killers in modern America. Hurricane Andrew did great physical damage, even though fatalities were few. What are sensible expenditures for research in meteorology and seismology, for mandatory enhancement of building codes and redevelopment, and for other measures that would mitigate natural disasters?

- The 9/11 fatalities are just 1.5 percent of those in the nation's worst epidemic (half a million died from flu in 1918) and also just 1.5 percent of the *annual* U.S. cancer fatalities. We have waged a "war" on cancer, at the expense of research on other less feared but deadly diseases; this war's success is equivocal (five-year survivability after detection is up, but so are cancer death rates — though mainly due to decades-old changes in smoking habits). Where should "homeland security" expenditures rank against medical expenditures?
- Impacts by kilometer-sized asteroids are extremely rare, but one could send civilization into a new Dark Age. The annualized American fatality rate is about 5 percent of the WTC fatalities, although such a cosmic impact has only 1/100th of a 1 percent chance of happening during the twenty-first century. Just a couple million dollars are now spent annually to search for threatening asteroids. Should we spend many billions to build a planetary defense shield, which would statistically be in proportion to what we now spend on homeland security and the war on terrorism? Might the threat to our civilization's very existence raise the stakes above even the terrorist threat?

To us, these comparisons suggest that the nation's post-9/11 expenditures have been lopsidedly large, and that a balanced approach would "give back" some funds to reduce deaths from falls, suicide, murder, highway accidents, natural disasters (including even asteroid impacts), malnutrition, and preventable or curable diseases . . . and give back our civil liberties and just the plain pleasures of life, such as the arts and humanities, exploration, and national parks. And if truly effective means to end wars could be found, they would be especially worthy of funds, given the death toll from twentieth-century wars. Before homeland security becomes dominated by vested bureaucracies and constituencies, there may yet be time to question its dominant role in our priorities.

We advocate shifting toward objective cost-benefit analyses and equitable evaluation of the relative costs of saving human lives. Of course, subjective judgments have some validity beyond strict adherence to numerical odds. But we need a national dialogue to address these issues dispassionately so that future governmental decisions can eschew immediate, impulsive reactions. Individual skeptics, in our own lives, can exemplify sensible choices. Among the many dumb things we should avoid (e.g., smoking, driving without a seatbelt, or letting kids play with firearms), we must also avoid driving instead of flying, acquiescing uncomplainingly to ineffective searches at local buildings and events, and generally yielding to the new "homeland security" mania. Clear thinking about risks, rather than saying that "any improvement in security is worth it," can reduce our societal vulnerability to terrorism.

One constructive antidote to post-9/11 trauma is to enhance the information available and to foster a sound appreciation, evaluation, and use of the information. Life is inherently risky, unpredictable, and subject to things we cannot know . . . but there are things we do know and can understand. Rather than scaring people about sharks, serial killers, and anthrax, the mass media could help people understand the real risks in their everyday environments and activities. Educational institutions should help students develop the critical skills necessary to make rational choices. While avoiding intrusions into personal liberties, government could nevertheless collect and assess statistical data in those arenas (like air travel) where potential dangers lurk, concentrating protective efforts and law enforcement where they are most efficacious.

To conclude, we suggest that most homeland security expenditures, which in the zero-sum budget game are diverted from other vital purposes, are terribly expensive and disproportionate to competing needs for preventing other causes of death and misery in our society. While prudent, focused improvements in security are called for, the sheer cost of most security initiatives greatly distorts the way we address the many threats to our individual and collective well-being. Our greatest vulnerability to terrorism is the persisting, irrational fear of terrorism that has gripped our country. We must start behaving like the informed, reasoning beings we profess to be.

DAVID EWING DUNCAN

DNA as Destiny

FROM *Wired*

I FEEL NAKED. Exposed. As if my skin, bone, muscle tissue, cells, have all been peeled back, down to a tidy swirl of DNA. It's the basic stuff of life, the billions of nucleotides that keep me breathing, walking, craving, and just being. Eight hours ago, I gave a few cells, swabbed from inside my cheek, to a team of geneticists. They've spent the day extracting DNA and checking it for dozens of hidden diseases. Eventually, I will be tested for hundreds more. They include, as I will discover, a nucleic time bomb ticking inside my chromosomes that might one day kill me.

For now I remain blissfully ignorant, awaiting the results in an office at Sequenom, one of scores of biotech startups incubating in the canyons north of San Diego. I'm waiting to find out if I have a genetic proclivity for cancer, cardiac disease, deafness, Alzheimer's, or schizophrenia.

This, I'm told, is the first time a healthy human has ever been screened for the full gamut of genetic disease markers. Everyone has errors in his or her DNA, glitches that may trigger a heart spasm or cause a brain tumor. I'm here to learn mine.

Waiting, I wonder if I carry some sort of Pandora gene, a hereditary predisposition to peek into places I shouldn't. Morbid curiosity is an occupational hazard for a writer, I suppose, but I've never been bothered by it before. Yet now I find myself growing nervous and slightly flushed. I can feel my pulse rising, a cardiovascular response that I will soon discover has, for me, dire implications.

In the coming days, I'll seek a second opinion, of sorts. Curious about where my genes come from, I'll travel to Oxford and visit an "ancestral geneticist" who has agreed to examine my DNA for links back to progenitors whose mutations have been passed on to me. He will reveal the seeds of my individuality and the roots of the diseases that may kill me — and my children.

For now, I wait in an office at Sequenom, a sneak preview of a trip to the DNA doctor, circa 2008. The personalized medicine being pioneered here and elsewhere prefigures a day when everyone's genome will be deposited on a chip or stored on a gene card tucked into a wallet. Physicians will forecast illnesses and prescribe preventive drugs custom-fitted to a patient's DNA, rather than the one-size-fits-all pharmaceuticals that people take today. Gene cards might also be used to find that best-suited career, or a DNA-compatible mate, or, more darkly, to deny someone jobs, dates, and meds because their nucleotides don't measure up. It's a scenario Andrew Niccol imagined in his 1997 film, *Gattaca,* where embryos in a not-too-distant future are bioengineered for perfection and where genism — discrimination based on one's DNA — condemns the lesser-gened to scrubbing toilets.

The *Gattaca*-like engineering of defect-free embryos is at least twenty or thirty years away, but Sequenom and others plan to take DNA testing to the masses in just a year or two. The prize: a projected $5 billion market for personalized medicine by 2006 and billions, possibly hundreds of billions, more for those companies that can translate the errors in my genome and yours into custom pharmaceuticals.

Sitting across from me is the man responsible for my gene scan: Andi Braun, chief medical officer at Sequenom. Tall and sinewy, with a long neck, glasses, and short gray hair, Braun, forty-six, is both jovial and German. Genetic tests are already publicly available for Huntington's disease and cystic fibrosis, but Braun points out that these illnesses are relatively rare. "We are targeting diseases that impact millions," he says in a deep Bavarian accent, envisioning a day when genetic kits that can assay the whole range of human misery will be available at Wal-Mart, as easy to use as a home pregnancy test.

But a kit won't tell me if I'll definitely get a disease, just if I have a bum gene. What Sequenom and others are working toward is pinning down the probability that, for example, a colon cancer gene will actually trigger a tumor. To know this, Braun must analyze the DNA of thousands of people and tally how many have the colon cancer gene, how many actually get the disease, and how many don't. Once these data are gathered and crunched, Braun will be able to tell you, for instance, that if you have the defective gene, you have a 40 percent chance, or maybe a 75 percent chance, of getting the disease by age fifty, or ninety. Environmental factors such as eating right — or wrong — and smoking also weigh in. "It's a little like predicting the weather," says Charles Cantor, the company's cofounder and chief scientific officer.

Braun tells me that, for now, his tests offer only a rough sketch of my genetic future. "We can't yet test for everything, and some of the information is only partially understood," he says. It's a peek more through a rudimentary eyeglass than a Hubble Space Telescope. Yet I will be able to glimpse some of the internal programming bequeathed to me by evolution and that I, in turn, have bequeathed to my children — Sander, Danielle, and Alex, ages fifteen, thirteen, and seven. They are a part of this story, too. Here's where I squirm, because as a father I pass on not only the ingredients of life to my children but the secret codes of their demise — just as I have passed on my blue eyes and a flip in my left brow that my grandmother called "a little lick from God." DNA is not only the book of life, it is also the book of death, says Braun: "We're all going to die, *ja?*"

Strictly speaking, Braun is not looking for entire genes, the long strings of nucleotides that instruct the body to grow a tooth or create white blood cells to attack an incoming virus. He's after single nucleotide polymorphisms, or SNPs (pronounced "snips"), the tiny genetic variations that account for nearly all differences in humans.

Imagine DNA as a ladder made of rungs — 3 billion in all — spiraling upward in a double helix. Each step is a base pair, designated by two letters from the nucleotide alphabet of G, T, A, and C. More than 99 percent of these base pairs are identical in all humans, with only about one in a thousand SNPs diverging to make us distinct. For instance, you might have a CG that makes you susceptible to diabetes, and I might have a CC, which makes it far less likely I will get this disease.

This is all fairly well known: Genetics 101. What's new is how startups like Sequenom have industrialized the SNP identification process. Andi Braun and Charles Cantor are finding thousands of new SNPs a day, at a cost of about a penny each.

Braun tells me that there are possibly a million SNPs in each person, though only a small fraction are tightly linked with common ailments. These disease-causing SNPs are fueling a biotech bonanza; the hope is that after finding them, the discoverers can design wonder drugs. In the crowded SNP field, Sequenom vies with Iceland-based deCode Genetics and American companies such as Millennium Pharmaceuticals, Orchid BioSciences, and Celera Genomics, as well as multinationals like Eli Lilly and Roche Diagnostics. "It's the Oklahoma Land Grab right now," says Toni Schuh, Sequenom's CEO.

The sun sets outside Braun's office as my results arrive, splayed across his computer screen like tarot cards. I'm trying to maintain a steely, reportorial facade, but my heart continues to race.

Names of SNPs pop up on the screen: connexin 26, implicated in hearing loss; factor V leiden, which causes blood clots; and alpha-1 antitrypsin deficiency, linked to lung and liver disease. Beside each SNP are codes that mean nothing to me: 13q11-q12, 1q23, 14q32.1. Braun explains that these are addresses on the human genome, the P.O. box numbers of life. For instance, 1q23 is the address for a mutant gene that causes vessels to shrink and impede the flow of blood — it's on chromosome 1. Thankfully, my result is negative. "So, David, you will not get the varicose veins. That's good, *ja?*" says Braun. One gene down, dozens to go.

Next up is the hemochromatosis gene. This causes one's blood to retain too much iron, which can damage the liver. As Braun explains it, somewhere in the past, an isolated human community lived in an area where the food was poor in iron. Those who developed a mutation that stores high levels of iron survived, and those who didn't became anemic and died, failing to reproduce. However, in these iron-rich times, hemochromatosis is a liability. Today's treatment? Regular bleeding. "You tested negative for this mutation," says Braun. "You do not have to be bled."

I'm also clean for cystic fibrosis and for a SNP connected to lung cancer.

Then comes the bad news. A line of results on Braun's monitor shows up red and is marked "MT," for mutant type. My body's programming code is faulty. There's a glitch in my system. Named ACE (for angiotensin-I converting enzyme), this SNP means my body makes an enzyme that keeps my blood pressure spiked. In plain English, I'm a heart attack risk.

My face drains of color as the news sinks in. I'm not only defective, but down the road, every time I get anxious about my condition, I'll know that I have a much higher chance of dropping dead. I shouldn't be surprised, since I'm told everyone has some sort of disease-causing mutation. Yet I realize that my decision to take a comprehensive DNA test has been based on the rather ridiculous assumption that I would come out of this with a clean genetic bill of health. I almost never get sick, and, at age forty-four, I seldom think about my physical limitations or death. This attitude is buttressed by a family largely untouched by disease. The women routinely thrive into their late eighties and nineties. One great-aunt lived to age one hundred and one; she used to bake me cupcakes in her retirement home when I was a boy. And some of the Duncan menfolk are pushing ninety-plus. My parents, now entering their seventies, are healthy. In a flash of red MTs, I'm glimpsing my own future, my own mortality. I'm slated to keel over, both hands clutching at my heart.

"Do you have any history in your family of high blood pressure or heart disease?" asks Matthew McGinniss, a Sequenom geneticist standing at Braun's side.

"No," I answer, trying to will the color back into my face. Then a second MT pops up on the screen — another high blood pressure mutation. My other cardiac indicators are OK, which is relatively good news, though I'm hardly listening now. I'm already planning a full-scale assault to learn everything I can about fighting heart disease — until McGinniss delivers an unexpected pronouncement. "These mutations are probably irrelevant," he says. Braun agrees: "It's likely that you carry a gene that keeps these faulty ones from causing you trouble — DNA that we have not yet discovered."

The SNPs keep rolling past, revealing more mutations, including a type 2 diabetes susceptibility, which tells me I may want to steer clear of junk food. More bad news: I don't have a SNP called CCR5 that prevents me from acquiring HIV, nor one that seems to shield smokers from lung cancer. "*Ja,* that's my favorite," says Braun, himself a smoker. "I wonder what Philip Morris would pay for that."

By the time I get home, I realize that all I've really learned is, I might get heart disease, and I could get diabetes. And I should avoid smoking and unsafe sex — as if I didn't already know this. Obviously, I'll now watch my blood pressure, exercise more, and lay off the Cap'n Crunch. But beyond this, I have no idea what to make of the message Andi Braun has divined from a trace of my spit.

Looking for guidance, I visit Ann Walker, director of the Graduate Program for Genetic Counseling at the University of California at Irvine. Walker explains the whats and hows, and the pros and cons, of DNA testing to patients facing hereditary disease, pregnant couples concerned with prenatal disorders, and anyone else contemplating genetic evaluation. It's a tricky job because, as I've learned, genetic data are seldom clear-cut.

Take breast cancer, Walker says. A woman testing positive for BRCA1, the main breast cancer gene, has an 85 percent chance of actually getting the cancer by age seventy, a wrenching situation, since the most effective method of prevention is a double mastectomy. What if a woman has the operation and it turns out she's among those 15 percent who carry the mutation but will never get the cancer? Not surprisingly, one study, conducted in Holland, found that half of the healthy women whose mothers developed breast cancer opt not to be tested for the gene, preferring ignorance and closer monitoring. Another example is the test for APoE, the Alzheimer's gene. Since the affliction has no cure, most people don't want to know their status. But some do. A positive result, says Walker, allows them to put their affairs in order and prepare for their own dotage. Still, the news can be devastating. One biotech executive told me that a cousin of his committed suicide when he tested positive for Huntington's, having seen the disease slowly destroy his father.

Walker pulls out a chart and asks about my family's medical details, starting with my grandparents and their brothers and sisters: what they suffered and died from, and when. My Texas grandmother died at ninety-two after a series of strokes. My ninety-one-year-old Missouri grandmom was headed to a vacation in Mexico with her eighty-eight-year-old second husband when she got her death sentence — ovarian cancer. The men died younger: my grandfathers in their late sixties, though they both have brothers still alive and healthy in their nineties. To the mix, Walker adds my parents and their siblings, all of whom are alive and healthy in their sixties and seventies; then my generation; and finally our children. She looks up and smiles: "This is a pretty healthy group."

Normally, Walker says, she would send me home. Yet I'm sitting across from her, not because my parents carry some perilous SNP, but as a healthy man who is after a forecast of future maladies. "We have no real training yet for this," she says, and tells me the two general rules of genetic counseling: No one should be screened unless there is an effective treatment or readily available counseling; and the information should not bewilder people or present them with unnecessary trauma.

Many worry that these prime directives may be ignored by Sequenom and other startups that need to launch products to survive. FDA testing for new drugs can take up to ten years, and many biotech firms feel pressure to sell something in the interim. "Most of these companies need revenue," says the University of Pennsylvania's Arthur Caplan, a top bioethicist. "And the products they've got now are diagnostic. Whether they are good ones, useful ones, necessary ones, accurate ones, seems less of a concern than that they be sold." Caplan also notes that the FDA does not regulate these tests. "If it was a birth control test, the FDA would be all over it."

I ask Caplan about the *Gattaca* scenario of genetic discrimination. Will a woman dump me if she finds out about my ACE? Will my insurance company hike my rate? "People are denied insurance and jobs right now," he says, citing sickle cell anemia, whose sufferers and carriers, mostly black, have faced job loss and discrimination. No federal laws exist to protect us from genism, or from insurers and employers finding out our genetic secrets. "Right now, you're likely going to be more disadvantaged than empowered by genetic testing," says Caplan.

After probing my genetic future, I jet to England to investigate my DNA past. Who are these people who have bequeathed me this tainted bloodline? From my grandfather Duncan, an avid genealogist, I already know that my paternal ancestors came from Perth, in south-central Scotland. We can trace the name back to an Anglican priest murdered in Glasgow in 1680 by a mob of Puritans. His six sons escaped and settled in Shippensburg, Pennsylvania, where their descendants lived until my great-great-grandfather moved west to Kansas City in the 1860s.

In an Oxford restaurant, over a lean steak and a heart-healthy merlot, I talk with geneticist Bryan Sykes, a linebacker-sized fifty-five-year-old with a baby face and an impish smile. He's a molecular biologist at the university's Institute of Molecular Medicine and the author of the best-selling *Seven Daughters of Eve*. Sykes first made headlines in 1994 when he used DNA to directly link a 5,000-year-old body discovered frozen and intact in an Austrian glacier to a twentieth-century Dorset woman named Marie Mosley. This stunning genetic connection between housewife and hunter-gatherer launched Sykes's career as a globe-trotting genetic gumshoe. In 1995, he confirmed that bones dug up near Ekaterinburg, Russia, were the remains of Czar Nicholas II and his family by comparing the body's DNA with that of the czar's living relatives, including Britain's Prince Philip. Sykes debunked explorer Thor Heyerdahl's *Kon-Tiki* theory by tracing Polynesian genes to Asia, not the Americas, and similarly put the lie to the *Clan of the Cave Bear* hypothesis, which held that the Neanderthal interbred with our ancestors, the Cro-Magnon, when the two subspecies coexisted in Europe 15,000 years ago.

Sykes explains to me that a bit of DNA called mtDNA is key to his investigations. A circular band of genes residing separately from the twenty-three chromosomes of the double helix, mtDNA is passed down solely through the maternal line. Sykes used mtDNA to discover something astounding: Nearly every European can be traced back to just seven women living 10,000 to 45,000 years ago. In his book, Sykes gives these seven ancestors hokey names and tells us where they most likely lived: Ursula, in Greece (circa 43,000 B.C.), and Velda, in northern Spain (circa 15,000 B.C.), to name two of the "seven daughters of Eve." (Eve was the ur-mother who lived 150,000 years ago in Africa.)

Sykes has taken swab samples from the cheeks of more than 10,000 people, charging $220 to individually determine a person's mtDNA type. "It's not serious genetics," Sykes admits, "but people like to know their roots. It makes genetics less scary and shows us that, through our genes, we are all very closely related." He recently expanded his tests to include non-Europeans. The Asian daughters of Eve are named Emiko, Nene, and Yumio, and their African sisters are Lamia, Latifa, and Ulla, among others.

Before heading to England, I had mailed Sykes a swab of my cheek cells. Over our desserts in Oxford he finally offers up the results. "You are descended from Helena," he pronounces. "She's the most common daughter of Eve, accounting for some 40 percent of Europeans." He hands me a colorful certificate, signed by him, that heralds my many-times-great-grandma and tells me that she lived 20,000 years ago in the Dordogne Valley of France. More interesting is the string of genetic letters from my mtDNA readout that indicate I'm mostly Celtic, which makes sense. But other bits of code reveal traces of Southeast Asian DNA, and even a smidgen of Native American and African.

This doesn't quite have the impact of discovering that I'm likely to die of a heart attack. Nor am I surprised about the African and Indian DNA, since my mother's family has lived in the American South since the seventeenth century. But Southeast Asian? Sykes laughs. "We are all mutts," he says. "There is no ethnic purity. Somewhere over the years, one of the thousands of ancestors who contributed to your DNA had a child with someone from Southeast Asia." He tells me a story about a blond, blue-eyed surfer from Southern California who went to Hawaii to apply for monies awarded only to those who could prove native Hawaiian descent. The grant-givers laughed — until his DNA turned up traces of Hawaiian.

The next day, in Sykes's lab, we have one more test: running another ancestry marker in my Y chromosome through a database of 10,000 other Ys to see which profile is closest to mine. If my father was in the database, his Y chromosome would be identical, or possibly one small mutation off. A cousin might deviate by one tick. Someone descended from my native county of Perth might be two or three mutations removed, indicating that we share a common ancestor hundreds of years ago. Sykes tells me these comparisons are used routinely in paternity cases. He has another application. He is building up Y-chromosome profiles of surnames: men with the same last name whose DNA confirms that they are related to common ancestors.

After entering my mtDNA code into his laptop, Sykes looks intrigued, then surprised, and suddenly moves to the edge of his seat.

Excited, he reports that the closest match is, incredibly, him —
Bryan Sykes! "This has never happened," he says, telling me that I
am a mere one mutation removed from him, and two from the average profile of a Sykes. He has not collected DNA from many
other Duncans, he says, though it appears as if sometime in the
past 400 years a Sykes must have ventured into Perth and then had
a child with a Duncan. "That makes us not-so-distant cousins," he
says. We check a map of Britain on his wall, and sure enough, the
Sykes family's homeland of Yorkshire is less than 200 miles south of
Perth.

The fact that Sykes and I are members of the same extended
family is just a bizarre coincidence, but it points to applications beyond simple genealogy. "I've been approached by the police to use
my surnames data to match up with DNA from an unknown suspect found at a crime scene," says Sykes. Distinctive genetic markers can be found at the roots of many family trees. "This is possible,
to narrow down a pool of suspects to a few likely surnames. But it's
not nearly ready yet."

Back home in California, I'm sweating on a StairMaster at the gym,
wondering about my heart. I wrap my hands around the grips and
check my pulse: 129. Normal. I pump harder and top out at 158.
Also normal. I think about my visit a few days earlier — prompted
by my gene scan — to Robert Superko, a cardiologist. After per-
forming another battery of tests, he gave me the all clear — except
for one thing. Apparently, I have yet another lame heart gene, the
atherosclerosis susceptibility gene ATHS, a SNP that causes plaque
in my cardiac bloodstream to build up if I don't exercise far more
than average — which I do, these days, as a slightly obsessed biker
and runner. "As long as you exercise, you'll be fine," Superko advised, a bizarre kind of life sentence that means that I must pedal
and jog like a madman or face — what? A triple bypass?

Pumping on the StairMaster, I nudge the setting up a notch,
wishing, in a way, that I either knew for sure I was going to die on,
say, February 17, 2021, or that I hadn't been tested at all. As it is,
the knowledge that I have an ACE and ATHS deep inside me will
be nagging me every time I get short of breath.

The last results from my DNA workup have also come in. Andi
Braun has tested me for seventy-seven SNPs linked to lifespan in or-

der to assess when and how I might get sick and die. He has given me a score of .49 on his scale. It indicates a lifespan at least 20 percent longer than that of the average American male, who, statistically speaking, dies in his seventy-fourth year. I will likely live, then, to the age of eighty-eight. That's forty-four years of StairMaster to go.

Braun warns that this figure does not take into account the many thousands of other SNPs that affect my life, not to mention the possibility that a piano could fall on my head.

That night, I put my seven-year-old, Alex, to bed. His eyes droop under his bright white head of hair as I finish reading *Captain Underpants* aloud. Feeling his little heart beating as he lies next to me on his bed, I wonder what shockers await him inside his nucleotides, half of which I gave him. As I close the book and then sing him to sleep, I wonder if he has my culprit genes. I don't know, because he hasn't been scanned. For now, he and the rest of humanity are living in nearly the same blissful ignorance as Helena did in long-ago Dordogne. But I do know one thing: Alex has my eyebrow, the "lick of God." I touch his flip in the dark, and touch mine. He stirs, but it's not enough to wake him.

TIMOTHY FERRIS

Astronomy's New Stars

FROM *Smithsonian*

AT SUNDOWN, at a star party on the high Texas plains near Fort Davis, west of the Pecos, the parched landscape was crowded with telescopes. Reared against the darkening skies to the west rose a set of rolling foothills known jocularly as the Texas Alps. To the east of us lay dinosaur country, with its wealth of oil.

The stars came out with imposing clarity — Orion fleeing toward the western horizon, pursued by the dog star, brilliant white Sirius, the square of Corvus the crow to the southeast, the scythe of Leo the lion near the zenith. The planet Jupiter stood almost at the zenith; scores of telescopes were pointed toward it, like heliotropes following the Sun. As the gathering darkness swallowed up the valley, the sight of the observers was replaced by landbound constellations of ruby LED indicators on the telescopes' electronics, the play of red flashlights, and voices — groans, labored breathing, muttered curses, and sporadic cries of delight when a bright meteor streaked across the sky. Soon it was dark enough to see the zodiacal light — sunlight reflected off interplanetary dust grains ranging out past the asteroid belt — stabbing the western sky like a distant searchlight. When the Milky Way rose over the hills to the east, it was so bright that I at first mistook it for a bank of clouds. Under skies this transparent, the Earth becomes a perch, a platform from which to view the rest of the universe.

I had come here to observe with Barbara Wilson, legendary for her sharp-eyed pursuit of things dark and distant. I found her atop a small ladder, peering through her 20-inch Newtonian — an instrument tweaked and collimated to within an inch of its life, with eyepieces that she scrubs with Q-Tips before each observing session, using a mixture of Ivory soap, isopropyl alcohol, and distilled water. On an observing table, Barbara had set up *The Hubble Atlas of Galaxies,* the *Uranometria 2000* star atlas, a night-vision star chart illuminated from behind by a red-bulb light box, a laptop computer pressed into service as yet another star atlas, and a list of things she hoped to see. I'd never heard of most of the items on her list, much less seen them. They included Kowal's Object (which, Barbara informed me, is a dwarf galaxy in Sagittarius), the galaxy Molonglo-3, the light from which set out when the universe was half its present age, and obscure nebulae with names like Minkowski's Footprint, Red Rectangle, and Gomez's Hamburger.

"I'm looking for the jet in M87," Barbara called down to me from the ladder. M87 is a galaxy located near the center of the Virgo cluster, 60 million light-years from Earth. A white jet protrudes from its nucleus. It is composed of plasma — free atomic nuclei and electrons, the survivors of events sufficiently powerful to have torn atoms apart — spat out at nearly the velocity of light from near the poles of a massive black hole at the center of this giant elliptical galaxy. (Nothing can escape from inside a black hole, but its gravitational field can slingshot matter away at high speeds.) To study the structure of the jet to map dark clouds in M87, professional astronomers use the most powerful instruments available, including the Hubble Space Telescope. I'd never heard of an amateur's having seen it.

There was a long pause. Then Barbara exclaimed, "It's there! I mean, it's *so* there!"

She climbed down the ladder, her smile bobbing in the dark. "I saw it once before, from Columbus," she said, "but I couldn't get anybody to confirm it for me — couldn't find anyone who had the patience that it takes to see this thing. But it's so obvious once you see it that you just go, 'Wow!' Are you ready to try?"

I climbed the ladder, focused the eyepiece, and examined the softly glowing ball of M87, inflated like a blowfish at a magnification of 770x. No jet yet, so I went into standard dim-viewing practice. Relax, as in any sport. Breathe fairly deeply, to make sure the brain gets plenty of oxygen. Keep both eyes open, so as not to strain the muscles in the one you're using. Cover your left eye with your palm or just blank it out mentally — which is easier to do than it sounds — and concentrate on what you're seeing through the telescope. Check the chart to determine just where the object is in the field of view, then look a bit away from that point: the eye is more sensitive to dim light just off center than straight ahead. And, as Barbara says, be patient. Once, in India, I peered through a spotting telescope at a patch of deep grass for more than a minute before realizing that I was seeing the enormous orange and black head of a sleeping Bengal tiger. Stargazing is like that. You can't hurry it.

Then, suddenly, there it was — a thin, crooked, bone-white finger, colder and starker in color than the pewter starlight of the galaxy itself, against which it now stood out. How wonderful to see something so grand, after years of admiring its photographs. I came down the ladder with a big smile of my own. Barbara called a coffee break and her colleagues departed for the ranch house cafeteria, but she remained by the telescope in case anyone else came along who might want to see the jet in M87.

Amateur astronomy had gone through a revolution since I started stargazing in the 1950s. Back then, most amateurs used reedy telescopes like my 2.4-inch refractor. A 12-inch reflector was considered a behemoth, something you told stories about should you be lucky enough to get a look through one. Limited by the light-gathering power of their instruments, amateurs mostly observed bright objects, like the craters of the Moon, the satellites of Jupiter, the rings of Saturn, along with a smattering of prominent nebulae and star clusters. If they probed beyond the Milky Way to try their hand at a few nearby galaxies, they saw little more than dim gray smudges.

Professional astronomers, meanwhile, had access to big West Coast telescopes like the legendary 200-inch at Palomar Mountain in Southern California. Armed with the most advanced technology of the day and their own rigorous training, the professionals got results. At Mount Wilson Observatory, near Pasadena, the astronomer Harlow Shapley in 1918–19 established that the Sun is located toward one edge of our galaxy, and Edwin Hubble in 1929 determined that the galaxies are being carried apart from one another with the expansion of cosmic space. Professionals like these became celebrities, lionized in the press as hawkeyed lookouts probing the mysteries of deep space.

Which, pretty much, they were: theirs was a golden age, when our long-slumbering species first opened its eyes to the universe beyond its home galaxy. But observing the professional way wasn't usually a lot of fun. To be up there in the cold and the dark, riding in the observer's cage and carefully guiding a long exposure on a big glass photographic plate, with icy stars shining through the dome slit above and starlight puddling below in a mirror the size of a trout pond, was indubitably romantic but also a bit nerve-racking. Big-telescope observing was like making love to a glamorous movie star: you were alert to the honor of the thing, but aware that plenty of suitors were eager to take over should your performance falter.

Nor did academic territoriality, jealous referees, and the constant competition for telescope time make professional astronomy a day at the beach. As a brilliant young cosmologist once told me, "A career in astronomy is a great way to screw up a lovely hobby."

So it went, for decades. Professionals observed big things far away and published in the prestigious *Astrophysical Journal* — which, as if to rub it in, ranked papers by the distances of their subjects, with galaxies at the front of each issue, stars in the middle, and planets, on the rare occasion that they appeared in the *Journal* at all, relegated to the rear. Amateurs showed schoolchildren the rings of Saturn at 76 power through a tripod-mounted spyglass at the State Fair. Inevitably, a few professionals disdained the amateurs. When Clyde Tombaugh discovered Pluto, the astronomer Joel Stebbins, usually a more charitable man, dismissed him as "a sub-amateur assistant." There were of course professionals who kept up good relationships with amateurs, and amateurs who did solid work without fretting over their status. But generally speaking, the amateurs lived in the valley of the shadow of the mountaintops. Which was odd, in a way, because for most of its long history, astronomy has been primarily an amateur pursuit.

The foundations of modern astronomy were laid largely by amateurs. Nicolaus Copernicus, who in 1543 moved the Earth from the center of the universe and put the Sun there instead (thus replacing a dead-end mistake with an open-ended mistake, one that encouraged the raising of new questions), was a Renaissance man, adept at many things, but only a sometime astronomer. Johannes Kepler, who discovered that planets orbit in ellipses rather than circles, made a living mainly by casting horoscopes, teaching grade school, and scrounging royal commissions to support the publication of his books. Edmond Halley, after whom the comet is named, was an amateur whose accomplishments — among them a year spent observing from St. Helena, a South Atlantic island so remote that Napoleon Bonaparte was sent there to serve out his second and terminal exile — got him named Astronomer Royal.

Even in the twentieth century, while they were being eclipsed by the burgeoning professional class, amateurs continued to make valuable contributions to astronomical research. Arthur Stanley Williams, a lawyer, charted the differential rotation of Jupiter's clouds and created the system of Jovian nomenclature used in Jupiter studies ever since. Milton Humason, a former watermelon farmer who worked as a muleteer at Mount Wilson, teamed up with the astronomer Edwin Hubble to chart the size and expansion rate of the universe.

The solar research conducted by the industrial engineer Robert McMath, at an observatory he built in the rear garden of his home in Detroit, so impressed astronomers that he was named to the National Academy of Sciences, served as president of the American Astronomical Society, a professional organization, and helped plan Kitt Peak National Observatory in Arizona, where the world's largest solar telescope was named in his honor.

Why were the amateurs, having played such important roles in astronomy, eventually overshadowed by the professionals? Because astronomy, like all the sciences, is young — less than 400 years old, as a going concern — and somebody had to *get* it going. Its instigators could not very well hold degrees in fields that didn't yet exist. Instead, they had to be either professionals in some related field, such as mathematics, or amateurs doing astronomy for the love of it. What counted was competence, not credentials.

Amateurs, however, were back on the playing field by about 1980. A century of professional research had greatly increased the range of observational astronomy, creating more places at the table than there were professionals to fill them. Meanwhile, the ranks of amateur astronomy had grown, too, along with the ability of the best amateurs to take on professional projects and also to pursue innovative research. "There will always remain a division of labor between professionals and amateurs," wrote the historian of science John Lankford in 1988, but "it may be more difficult to tell the two groups apart in the future."

The amateur astronomy revolution was incited by three technological innovations — the Dobsonian telescope, CCD light-sensing devices, and the Internet. Dobsonians are reflecting telescopes constructed from cheap materials. They were invented by John Dobson, a populist proselytizer who championed the view that the worth of telescopes should be measured by the number of people who get to look through them.

Dobson was well known in San Francisco as a spare, ebullient figure who would set up a battered telescope on the sidewalk, call out to passersby to "Come see Saturn!" or "Come see the Moon!" then whisper astronomical lore in their ears while they peered into the eyepiece. To the casual beneficiaries of his ministrations, he came off as an aging hippie with a ponytail, a ready spiel, and a gaudily painted telescope so dinged up that it looked as if it had been dragged behind a truck. But astronomical sophisticates came to recognize his telescopes as the carbines of a scientific revolution. Dobsonians employed the same simple design that Isaac Newton dreamed up when he wanted to study the great comet of 1680 — a tube with a concave mirror at the bottom to gather starlight and a small, flat, secondary mirror near the top to bounce the light out to an eyepiece on the side — but they were made from such inexpensive materials that you could build or buy a big Dobsonian for the cost of a small traditional reflector. You couldn't buy a Dobsonian from John Dobson, though; he refused to profit from his innovation.

Observers armed with big Dobsonians didn't have to content themselves with looking at planets and nearby nebulae: they could explore thousands of galaxies, invading deep-space precincts previously reserved for the professionals. Soon, the star parties where amateur astronomers congregate were dotted with Dobsonians that towered 20 feet and more into the darkness. Now, thanks to Dobson, the greatest physical risk to amateur observers became that of falling from a rickety ladder high in the dark while peering through a gigantic Dobsonian. I talked with one stargazer whose Dobsonian stood so tall that he had to use binoculars to see the display on his laptop computer from atop the 15-foot ladder required to reach the eyepiece in order to tell where the telescope was pointing. He said he found it frightening to climb the ladder by day but forgot about the danger when observing by night. "About a third of the galaxies I see aren't catalogued yet," he mused.

Meanwhile the CCD had come along — the "charge-coupled device" — a light-sensitive chip that can record faint starlight much faster than could the photographic emulsions that CCDs soon began replacing. CCDs initially were expensive, but their price fell steeply. Amateurs who attached CCDs to large Dobsonians found themselves in command of light-gathering capacities comparable to that of the 200-inch Hale telescope at Palomar in the pre-CCD era.

The sensitivity of CCDs did not in itself do much to close the gap separating amateur from professional astronomers — since the professionals had CCDs too — but the growing quantity of CCDs in amateur hands vastly increased the number of telescopes on Earth capable of probing deep space. It was as if the planet had suddenly grown thousands of new eyes, with which it became possible to monitor many more astronomical events than there were professionals enough to cover. And, because each light-sensitive dot (or "pixel") on a CCD chip reports its individual value to the computer that displays the image it has captured, the stargazer using it has a quantitative digital record that can be employed to do photometry, as in measuring the changing brightness of variable stars.

Which brings us to the Internet. It used to be that an amateur who discovered a comet or an erupting star would dispatch a telegram to the Harvard College Observatory, from which a professional, if the finding checked out, sent postcards and telegrams to paying subscribers at observatories around the world. The Internet opened up alternative routes. Now an amateur who made a discovery — or thought he did — could send CCD images of it to other observers, anywhere in the world, in minutes. Global research networks sprang up, linking amateur and professional observers with a common interest in flare stars, comets, or asteroids. Professionals sometimes learned of new developments in the sky more quickly from amateur news than if they had waited for word through official channels, and so were able to study them more promptly.

If the growing number of telescopes out there gave the Earth new eyes, the Internet fashioned for it a set of optic nerves, through which flowed (along with reams of financial data, gigabytes of gossip, and cornucopias of pornography) news and images of storms raging on Saturn and stars exploding in distant galaxies. Amateur superstars emerged, armed with the skills, tools, and dedication to do what the eminent observational cosmologist Allan Sandage called "absolutely serious astronomical work." Some chronicled the weather on Jupiter and Mars, producing planetary images that rivaled those of the professionals in quality and surpassed them in documenting long-term planetary phenomena. Others monitored variable stars useful in determining the distances of star clusters and galaxies.

Amateurs discovered comets and asteroids, contributing to the continuing effort to identify objects that may one day collide with the Earth and that, if they can be found early enough, might be deflected to prevent such a catastrophe. Amateur radio astronomers recorded the outcries of colliding galaxies, chronicled the ionized trails of meteors falling in daytime, and listened for signals from alien civilizations.

The amateur approach had its limitations. Amateurs insufficiently tutored in the scientific literature sometimes acquired accurate data but did not know how to make sense of it. Those who sought to overcome their lack of expertise by collaborating with professionals sometimes complained that they wound up doing most of the work while their more prestigious partners got most of the credit. Others burned out, becoming so immersed in their hobby that they ran low on time, money, or enthusiasm and called it quits. But many amateurs enjoyed fruitful collaborations, and all were brought closer to the stars.

I met Stephen James O'Meara at the Winter Star Party, held annually alongside a sandy beach in West Summerland Key, Florida. Arriving after dark, I was greeted at the gate by Tippy D'Auria, the founder of the Winter Star Party, who led me through thickets of telescopes reared against the stars.

"Steve's up there, drawing Jupiter through my telescope," Tippy said, nodding toward the silhouette of a young man perched atop a stepladder at the eyepiece of a big Newtonian that was pointing into the southwest sky. Comfortable in my lawn chair, I listened to the elders talk — a mix of astronomical expertise and self-deprecatory wit, the antithesis of pomp — and watched O'Meara drawing. He would peer at length through the eyepiece, then down at his sketch pad and draw a line or two, then return to the eyepiece. It was the sort of work astronomers did generations ago, when observing could mean spending a night making one drawing of one planet.

O'Meara likes to describe himself as "a nineteenth-century observer in the twenty-first century," and in meeting him I hoped to better understand how someone who works the old-fashioned way, relying on his eye at the telescope rather than a camera or a CCD, had been able to pull off some of the most impressive observing feats of his time.

While still a teenager, O'Meara saw and mapped radial "spokes" on Saturn's rings that professional astronomers dismissed as illusory — until *Voyager* reached Saturn and confirmed that the spokes were real. He determined the rotation rate of the planet Uranus, obtaining a value wildly at variance with those produced by professionals with larger telescopes and sophisticated detectors, and proved to be right about that too. He was the first human to see Halley's comet on its 1985 return, a feat he accomplished using a 24-inch telescope at an altitude of 14,000 feet while breathing bottled oxygen.

After nearly an hour, O'Meara came down the ladder and made a gift of his drawing to Tippy, who introduced us. Clear-eyed, fit, and handsome, with black hair, a neatly trimmed beard, and a wide smile, O'Meara was dressed in a billowing white shirt and black peg pants. We repaired to the red-lit canteen for a cup of coffee and a talk.

Steve told me that he'd grown up in Cambridge, Massachusetts, the son of a lobster fisherman, and that his first childhood memory was of sitting in his mother's lap and watching the ruddy lunar eclipse of 1960. "From the very beginning I had an affinity with the sky," he said. "I just loved starlight." When he was about six years old he cut out a planisphere — a flat oval sky map — from the back of a box of corn flakes, and with it learned the constellations. "Even the tough kids in the neighborhood would ask me questions about the sky," he recalled. "The sky produced a wonderment in them. I believe that if inner-city kids had the opportunity to see the real night sky, they could believe in something greater than themselves — something that they can't touch, control, or destroy."

When O'Meara was about fourteen years old he was taken to a public night at the Harvard College Observatory, where he waited in line for a look through its venerable Clark 9-inch refractor. "Nothing happened for a long time," he recalled. "Eventually people started wandering off, discouraged. The next thing I knew I was inside the dome. I could hear a whirring sound and see the telescope pointing up at the stars, and a poor guy down there at the eyepiece — searching, searching — and he was sweating. I realized that he was trying to find the Andromeda galaxy. I asked him, 'What are you looking for?'

"'A galaxy far away.'

"I waited a few minutes, then asked, 'Is it Andromeda?' There was a silence, and finally he said, 'Yeah, but it's difficult to get, very complicated.'

"'Can I try?'

"'Oh, no, it's a very sophisticated instrument.'

"I said, 'You know, nobody's behind me. I can get it for you in two seconds.' I got it in the field of view.

"Everyone who had waited in line got to see the Andromeda galaxy through the telescope, and after they left he said, 'Show me what you know.' He was just a graduate student, and he didn't really know the sky. I showed him around, acquainted him with Messier galaxies and all sorts of things. We stayed up till dawn. The next morning he took me to the business office and they gave me a key, saying that if I helped them out with open houses, in return I could use the scope anytime I wanted. So now I was a fourteen-year-old kid with a key to the Harvard College Observatory!"

For years thereafter the observatory was O'Meara's second home. After school he would work afternoons in a Cambridge pharmacy, then spend his nights at the telescope, patiently making drawings of comets and planets. "Why draw at the telescope? Because what you get on film and CCD does not capture the essence of what you see with the eye," he told me. "Everyone looks at the world in a different way, and I'm trying to capture what I see and encourage others to look, to learn, to grow and understand, to build an affinity with the sky.

"Anyone who wants to be a truly great observer should start with the planets, because that is where you learn patience. It's amazing what you can learn to see, given enough time. That's the most important and critical factor in observing — time, time, time — though you never see it in an equation."

In the mid-1970s, O'Meara studied the rings of Saturn at the behest of Fred Franklin, a Harvard planetary scientist. He began seeing radial, spokelike features on one of the rings. He included the spokes in the drawings that he would slip under Franklin's office door in the morning. Franklin referred O'Meara to Arthur Alexander's *The Planet Saturn*. There O'Meara learned that the nineteenth-century observer Eugene Antoniadi had seen similar radial features in another ring.

But the consensus among astronomers was that they must be an illusion, because the differential rotation rate of the rings — they consist of billions of particles of ice and stone, each a tiny satellite, and the inner ones orbit faster than the outer ones — would smear out any such features. O'Meara studied the spokes for four more years, determining that they rotated with a period of ten hours — which is the rotation period of the planet, but not of the rings. "I did not find one person, honestly, who ever supported me in this venture," O'Meara recalled.

Then, in 1979, the *Voyager 1* spacecraft, approaching Saturn, took images that showed the spokes. "It was an overpowering emotion, to have that vindication at last," O'Meara said.

I asked Steve about his determination of the rotation period of Uranus. This had long been unknown, since Uranus is remote — it never gets closer than 1.6 billion miles from Earth — and shrouded in almost featureless clouds. He told me that Brad Smith, the astronomer who headed the *Voyager* imaging team, "called me one day and said, 'OK, Mr. Visual Guy, *Voyager* is going to be at Uranus in a few years, and I'm trying to first obtain the rotation period for Uranus. Do you think you can do it visually?' I said, 'Well, I'll try.'" O'Meara first read up on the history of Uranus observations and then inspected the planet repeatedly, starting in June 1980. He saw nothing useful until one night in 1981, when two fantastically bright clouds appeared. "I followed them as they did a sort of dance over time, and from these observations, with some help, I determined where the pole was, modeled the planet, and got a rotation period for each cloud, averaging around 16.4 hours." This number was disturbingly discordant. Brad Smith, observing with a large telescope at Cerro Tololo Observatory in Chile, was getting a rotation period of 24 hours, and a group of professional astronomers at the University of Texas, using CCD imaging, were also getting 24 hours.

To test O'Meara's vision, Harvard astronomers mounted drawings on a building across campus and asked him to study them through the 9-inch telescope he had used as a teenager. Although others could see little, O'Meara accurately reproduced the draw-ings. Impressed, the astronomers vouched for his Uranus work, and his results were published by the International Astronomical Union, a professional group. When *Voyager* reached Uranus, it con-firmed that the planet's rotation period, at the latitude of the clouds O'Meara had seen, was within one-tenth of an hour of his value.

We finished our coffee and made ready to go back out into the darkness. "I've always been strictly a visual observer, researching the sky with an eye to finding something new there," O'Meara said.

"We're all star people, in the sense that we're all created from star stuff, so it's in our genes, so to speak, that we're curious about the stars. They represent an ultimate power, something we cannot physically grasp. When people ask, 'Why, God?' they don't look down at the ground. They look up at the sky."

IAN FRAZIER

Terminal Ice

FROM *Outside*

WE ARE MELTING, like the Wicked Witch of the West. Soon there will be nothing left of us but our hat. From Chile to Alaska to Norway to Tibet, glaciers are going in reverse. Artifacts buried since the Stone Age emerge intact from the ice; in British Columbia, sheep hunters passing a glacier find protruding from it a prehistoric man, preserved even to his skin, his leather food pouch, and his fur cloak. All across the north, permafrost stops being perma-. In the Antarctic, some penguin populations decline. In Hudson Bay, ice appears later in the year and leaves earlier, giving polar bears less time to go out on it and hunt seals, causing them to be 10 percent thinner than they were twenty years ago, causing them to get into more trouble in the Hudson Bay town of Churchill, where (as it happens) summers are now twice as long. One day in August of 2000 an icebreaker goes to the North Pole and finds, not ice, but open ocean. The news is no surprise to scientists, who knew that the remote Arctic in summer has lots of ice-free areas. For the rest of us, a disorienting adjustment of the geography of Christmas is required.

Globally, there's a persistent trickling as enormities of ice unfreeze. The Greenland ice sheet loses 13 trillion gallons of fresh water a year, contributing a measurable percentage to the world's annual sea-level rise. Every year, the level of the sea goes up about the thickness of a dime. Other meltwater, and the warming of the planet, which causes water to expand, contribute too. A dime's thickness a year doesn't worry most people, so long as it doesn't get worse, which most scientists don't think it necessarily will anytime soon, though who can say for sure? The first nation to ratify the Kyoto Protocol on Climate Change is the island of Fiji, one eye on the Pacific lapping at its toes.

And every year, first attracting notice in the seventies, picking up speed in the steamy eighties and steamier nineties, giant icebergs begin splashing into the news. Usually they arrive in single-column stories on an inside page: "An iceberg twice as big as Rhode Island has broken away from Antarctica and is drifting in the Ross Sea . . . It is about 25 miles wide and 98 miles long." "The largest iceberg in a decade has broken off an ice shelf in Antarctica . . . as if Delaware suddenly weighed anchor and put out to sea."

Over the years, a number of Rhode Islands and Delawares of ice, and even a Connecticut, drift into type and out again. The more notable ones are sometimes called "celebrity icebergs," and in the cold Southern Ocean (all the biggest icebergs are from Antarctica) an occasional berg has a longevity in the spotlight that a human celebrity could envy. Iceberg C-2 — as scientists labeled it — drifts for twelve years and 5,700 miles, nearly circumnavigating Antarctica, before breaking into pieces of non-newsworthy size.

Glaciologists say there's probably no connection between global climate change and the increase in the numbers of big Antarctic icebergs. They say the ice shelves at the edge of the continent, from which these icebergs come, have grown out and shrunk back countless times in the past. Our awareness of the icebergs has mainly to do with satellite technology that allows us to see them as we never could before. Still, when you've recently been through the hottest year of the past six centuries and suddenly there's a 2,700-square-mile iceberg on the loose — well, people talk.

In recent times I did a lot of reading about icebergs, some of it at the library of a western university where the air outdoors was so full of smoke from forest fires that people were going around in gas masks. To the old question of whether we will end in fire or ice, the answer now seems to be: both. Fire's photogenic, media-friendly qualities may cause us sometimes to overlook its counterpart and to forget the spectacular entrance ice made onto the modern apocalyptic scene just ninety years ago. Ice plus the *Titanic* spawned nightmares of disaster that never seem to fade. There was a song people used to sing about the *Titanic*, part of which went:

It was on her maiden trip
When an iceberg hit the ship . . .

Of course, the iceberg didn't hit the ship, but the other way around. So forcefully did the iceberg enter our consciousness, however, we assume it must have meant to. Looming unannounced from the North Atlantic on April 14, 1912, it crashed the swells' high-society ball, discomfiting Mrs. Astor, leaving its calling card in the form of a cascade of ice on the starboard well deck, slitting the hull fatally 20 feet below the waterline, and then disappearing into the night. In an instant this "Shape of Ice," as the poet Thomas Hardy called it, had become more famous than all the celebrities onboard. In its dreadful individuality, it had become The Iceberg.

An inflatable toy version of it is sitting on my desk. It came with a *Titanic* bath toy some friends gave my children a while ago. The inflatable iceberg is roughly pyramidal, with three peaks — two of them small and a larger one in the middle. Whoever designed the toy must have seen the widely published photograph supposedly showing The Iceberg, or perhaps saw a cinematic iceberg based on the photo. Hours after the sinking, observers on a German ship reported an iceberg of this shape near the scene, and one of them took the famous photo. Two weeks later, another transatlantic steamer said it saw a different-looking iceberg surrounded by deck chairs, cushions, and other debris in a location where The Iceberg could have drifted. As is the case with many suspects, no positive identification could ever be made.

Northern Hemisphere icebergs like The Iceberg melt quickly once they drift down into the Atlantic, with its warming Gulf Stream. Almost certainly, within a few weeks of shaking up the world, The Iceberg had disappeared. Its ephemerality has only increased its fame; solid matter for just a few historic moments, it continues indefinitely in imaginary realms — for example, as a spooky cameo in the top-grossing movie of all time. The message of The Iceberg, common wisdom has it, concerns the inscrutability of our fate and the vanity of human pride. But when I meditate on ice and icebergs, I wonder if The Iceberg's message might have been simpler than that. Maybe the news The Iceberg bore was more ancient, powerful, planetary, and climatic. Maybe The Iceberg's real message wasn't about us but about ice.

*

Icebergs are pieces of freshwater ice of a certain size floating in the ocean or (rarely) a lake. They come from glaciers and other ice masses. Because of the physics of ice when it piles up on land, it spreads and flows, and as it does its advancing edge often meets a body of water. When the ice continues to flow out over the water, chunks of it break off in a process called calving. Some of the faster-moving glaciers in Greenland calve an average of two or three times a day during the warmer months. Icebergs are not the same as sea ice. Sea ice is frozen saltwater, and when natural forces break it into pieces, the larger ones are called not icebergs but ice floes. Icebergs are denser and harder than sea ice. When icebergs are driven by wind or current, sea ice parts before them like turkey before an electric carving knife. In former times, sailing ships that got stuck in sea ice sometimes used to tie themselves to an iceberg and let it pull them through.

A piece of floating freshwater ice must be at least 50 feet long to qualify as an iceberg, according to authorities on the subject. If it's smaller — say, about the size of a grand piano — it's called a growler. If it's about the size of a cottage, it's a bergy bit. Crushed-up pieces of ice that result when parts of melting icebergs disintegrate and come falling down are called "slob ice" by mariners. Students of icebergs have divided them by shape into six categories: blocky, wedge, tabular, dome, pinnacle, and dry-dock. The last of these refers to icebergs with columnar sections flanking a water-level area in the middle, like high-rise apartment buildings around a swimming pool.

At the edges of Antarctica, where plains of ice spread across the ocean and float on it before breaking off, most of the icebergs are tabular — flat on top, horizontal in configuration. In the Northern Hemisphere, because of the thickness of glacial ice and the way it calves, most icebergs are of the more dramatically shaped kinds. Tabular icebergs tend to be stable in the water, and scientists sometimes land in helicopters on the bigger ones to study them. Northern Hemisphere icebergs, with their smaller size and gothic, irregular shapes, often grow frozen seawater on the bottom, lose above-water ice structure to melting, and suddenly capsize and roll. Venturing onto such icebergs is a terrible idea.

Antarctica has about 90 percent of all the ice in the world; most of the rest of it is in Greenland. Those two places produce most of the world's icebergs — about 100,000 a year from the first, about

10,000 to 15,000 from the other. Glaciers in Norway, Russia, and Alaska produce icebergs, too. The *Exxon Valdez* went aground in Alaska's Prince William Sound partly because it had changed course to avoid icebergs. Scientists have not been observing icebergs long enough to say if there are substantially more of them today. They know that the total mass of ice in Greenland has decreased at an accelerated rate in recent years. In Antarctica, because of its size and other factors, scientists still don't know whether the continent as a whole is losing ice or not.

Some people have jobs that involve thinking about ice and icebergs all day long. A while ago I went to the National Ice Center in Suitland, Maryland, and met a few of them. The Ice Center is affiliated with the National Oceanic and Atmospheric Administration, the Coast Guard, and the Navy. The offices of the Ice Center are in one of the many long, three-story government buildings with extra-large satellite dishes on their roofs in a fenced-in, campuslike setting just across the Potomac from Washington. Antennae poking up from behind clumps of trees add to the spy-thriller atmosphere. The Ice Center's supervisor, a lieutenant commander in the Navy — many of the people who work at the center are military personnel — introduced me to Judy Shaffier, an ice analyst. She is a slim woman in her mid-thirties with a shaped haircut and avid dark eyes accustomed to spotting almost invisible details. An ice analyst looks at satellite images of ocean ice on a computer screen, compares the images to other weather information, and figures out what they mean. "It's a great job, really neat to tell people about at a party or something," Shaffier said. "But it takes explaining. When I say I analyze ice, sometimes people don't get that I mean *ice*. They think it must be one of those government-agency acronyms."

Much of the Ice Center is closed to visitors. That's the part where it pursues its main purpose, which is to provide classified information on ice conditions to the military. For example, a nuclear submarine can break through ice 3 feet thick or less; the center can tell a submarine how close to surfaceable ice it is. Many countries — Japan, Denmark, Great Britain, Russia, France, Sweden — have ice-watching agencies similar to the Ice Center. Any country involved in global ocean shipping needs ice information sometimes.

Providing it to merchant ships, scientific expeditions, and the general public is the nonclassified part of the center's job.

Shaffier led me into a room with darkened windows and computers all along the walls. The glow of screens in the dimness lit the faces of ice analysts tapping on keyboards, summoning up satellite pictures of ice-covered oceans and seas. "I was trained as a meteorologist originally," Shaffier said. "I started analyzing ice in '94. Mostly what we do here is sea ice. Each of us has areas we concentrate on. Mine are the Yellow Sea, the Sea of Okhotsk, the Ross Sea, and the Sea of Japan. I feel like I really know the ice in those seas. Tracking icebergs is kind of secondary. We do it to keep everyone informed about possible dangers in the shipping lanes — in the Southern Hemisphere only, because another agency handles North Atlantic bergs. Also, I guess we do it for scientific and geographic reasons, or because icebergs are just interesting.

"For us to track an iceberg it must be at least 10 nautical miles long," she continued. "We label each one according to the quadrant of Antarctica where it broke off. The quadrants are A through D, and after the letter we add a number that's based on how many other bergs from that quadrant we've tracked since we started doing this back in 1976. The A quadrant, between 90 west longitude and zero, has shelves that calve big icebergs all the time, and we've tracked a lot of bergs from there. A-38 was a recent one. And if a berg breaks up into pieces, any piece that's bigger than 10 nautical miles gets its own label, like A-38A, A-38B, and so on."

The subject turned to giant "celebrity" icebergs and whether she had a favorite.

"In March of 2000 I was sitting at this computer," she said. "Another analyst, Mary Keller, was sitting at the next one, and suddenly she said, 'My God! It's a huge iceberg!' A huge piece had broken off a shelf in the B quadrant since we'd last checked, a day or so before. There had been no stress fractures visible in the ice sheet; the calving was completely unexpected. This iceberg was the fifteenth in B, so we labeled it B-15. I had never seen a berg that size. It was awesome — 158 by 20 nautical miles. After it broke off, it kind of ratcheted itself along the coast, sliding on each low tide, slowly moving from where it began, and in the process it eventually split into two pieces, one of about 100 miles long and another of about

80. B-15 was the most exciting iceberg I've watched since I've been here."

Shaffier and I spent hours looking at computer images that hop-scotched the icy places of the globe. Some of the pictures were visible-light photographs; some were made by infrared imagery that indicated different ice temperatures by color. Iceberg ice is 20 or 30 Celsius degrees colder than sea ice, old sea ice is a few degrees colder than new sea ice; in general, the colder the ice, the more difficult it is to navigate through. In passing, we checked up on B-15A — the giant was partly blocking the entry to West Antarctica's McMurdo Sound, apparently stuck on underwater rocks.

"Let's look at the Larsen Ice Shelf, or what's left of it," Shaffier said. "Did you hear what just happened to Larsen B? The Larsen Shelf is in a part of West Antarctica where local temperatures have gone up 4 or 5 degrees over the past decades, and a few years ago a big part of the shelf, Larsen A, disintegrated almost completely. People said that the rest of the shelf, Larsen B, would probably go in the next two years. Well, a few weeks ago it went. Over a thousand square miles of ice — *poof*. One day the shelf was there, the next day it started to break up, and 35 days later the satellite images showed nothing but dark water and white fragments where solid white ice used to be. In about a month this major geographic feature of Antarctica ceased to exist."

Later, a glaciologist I talked to explained how the breakup of Larsen B probably occurred: Higher surface temperatures created pools of meltwater that accumulated on the ice during the summer months until the water flowed into ice crevices; once in the crevices, the water became a hydraulic wedge, forcing its greater weight down through the ice and cracking it apart. When pieces of the ice broke off, they pushed over the pieces next to them, like books falling on a shelf. The process had a swiftness and a magnitude glaciologists had never seen before, and it created the largest movement of ice in a single event in recent times. Unlike the calving of giant icebergs, the Larsen Shelf breakup occurred because of a sharp rise in local temperatures, which scientists believe was almost certainly the result of global warming.

"Actually, I've never seen an iceberg except in pictures," Shaffier told me. "The only ice on my computer screen I've ever seen in person was ice on Lake Erie near Brook Park, Ohio, where I used to live. Next fall I might get a place on the supply ship going down to McMurdo. I'd love to do that. I want to see the ice up close, but even more I want to hear it. The images on my screen are silent, of course. But think about when an iceberg 200 miles long by 21 miles wide breaks off Antarctica. Think what that sound must be."

According to scientists, probably no one has ever heard that sound or been present when a giant Antarctic iceberg calved. Whether that event is even accompanied by sound audible to humans, no one can yet say. Instruments that listen for underwater oceanic sounds sometimes pick up vibrations like a cello bow across strings — only much lower, below human hearing — which are believed to come from the friction of giant icebergs, though none of the vibrations have yet been matched to a specific calving. When giant icebergs run into something, however — when they collide with the ocean bottom or the land — they cause seismic tremors that register at listening stations halfway around the world.

I wanted to see an iceberg myself. Like Judy Shaffier, I had never seen one live. I took a cab from my house in New Jersey to Newark Airport and flew to Halifax, Nova Scotia, one of those far northern airports where connecting passengers half-sleepwalk in a twilight of individual time zones. From Halifax I flew to St. John's, Newfoundland, on an uncrowded plane over the coastline and then across ocean that was mailbox blue. I had my face close to the window, craning my neck to scan. A finger of Newfoundland appeared directly below the plane. In the other direction, on the ocean far to the east, I saw a blip of white. It got bigger as the plane approached, and I could make out what seemed to be two white oil-storage tanks rising from the ocean's surface. They looked so plausible, I was sure that was what they were. At slightly closer range, the deception fell away, and I saw they were both parts of a single iceberg; and moreover, one of the dry-dock kind I'd read about. This sighting excited me beyond all measure. From home to iceberg was about two hours of flying time.

The plane banked to the west and descended to land at St. John's, and at about a thousand feet it came over a high bluff above a fjordlike little bay, and in the middle of it was a large, tent-shaped iceberg. The plane passed over the iceberg in a second and five minutes later was at the gate. I hurried to the luggage claim, got my bag, picked up my rental car, and drove back to the bay where the iceberg was. At a turnout by the road skirting the bay, I got out. The berg rode there, rotating slightly back and forth, about 200 yards away. Its top had a sort of spinal effect, with knobs in a curving row like vertebrae. Small waves broke around it, and it seemed to give off a mist.

Icebergs are really white. Usually you don't see this kind of white unless you've just been born or are about to die. It's a hazmat suit, medical lab, hospital white. There are some antiseptic-blue overtones to it, too, and a whole spectrum of greens where the berg descends into the depths out of sight. In these latitudes, sea and land and sky wear the colors of hand-knit Scottish sweaters: the taupes, the teals, the tans, the oyster grays. Surrounded by these muted shades, icebergs stand out like sore thumbs, if a sore thumb could gleam white and rise five stories above the ocean and float.

I drove hundreds of miles up and down the Newfoundland coast looking for icebergs. When I spotted an iceberg in close, run aground on a point or in a cove, I went toward it in the car as far as I could and then hiked the rest of the way. One such hike led through meadows, down forest trails, and across slippery shoreline rocks all inclined in the same direction. Finally I got to the iceberg, which looked somewhat like a jawbone. It even had tooth-shaped serrations in the right place as it seesawed chewingly in the waves. Gulls on fixed wings shot past the headland entanglements of weather-killed trees, clouds turned in huge pinwheels above, waves crashed, the iceberg chewed. I had forgotten to bring water and my mouth was dry.

Among the small channels and troughs in the rocks, iceberg fragments were washing back and forth. I leaned down and scooped out a flat, oval piece about ten inches long. The sea had rounded and smoothed the hard, clear ice like a sea-smoothed stone. In it were tiny bubbles of air that had been trapped among fallen snowflakes millennia ago; the air had eventually become bubbles as later snowfalls compressed the snow to ice. Bubbles in icebergs are what cause them to reflect white light. Almost certainly, this piece of ice originally was part of the Greenland ice cap. A glacier like the Jakobshavn Glacier on Greenland's western coast probably calved this ice into Baffin Bay, where it may have remained for a year or more until currents took it north and then south into the Labrador Current, which brought it here.

I licked the ice, bit off a piece. It broke sharply and satisfyingly, like good peanut brittle. At places the Greenland ice cap is two miles thick. Climatologists have taken core samples clear through it. Chemical analysis of the ice and the air bubbles in these cores provides a picture of climate and atmosphere during the past 110,000 years. For a period covering all of recorded human history, the Greenland ice timeline is so exact that scientists can identify specific events with strata in the core sample — the year Vesuvius buried Pompeii, say, marked by chemical remnants of the Vesuvian eruption. In ice core samples dating from the Golden Age of Greece, they've found trace amounts of lead, dispersed into the atmosphere by early smelting processes and carried to Greenland on the winds. Lead traces in the ice increase slowly from Greek and Roman times, stay at about the same level during the Middle Ages, go up a lot after the beginning of the Industrial Revolution, crest during the twentieth century, with its leaded automobile fuels, and drop way off after the introduction of unleaded gasoline. Chemicals and other residues in the Greenland ice cores told scientists about temperatures, droughts, the coming and going of ice ages — more about paleoclimates than they had ever known. In particular, they showed how unstable global climate has been, how abruptly it sometimes changed, and how oddly mild and temperate were the recent few thousand years in which people developed civilization.

Scientifically useful concentrations of chemicals don't affect the purity of Greenland ice. Devices that measure impurities in parts per million usually register none in iceberg ice; to find substances other than air and water in it, measurement must be in smaller concentrations, like parts per billion or per trillion. As I sucked the iceberg piece, contemplating its ancientness, trying to taste the armor of Caesar or the ash of Krakatoa in infinitesimal traces, the pristine cold water seemed to evaporate through my membranes with no intervening stage. Before I finished the fragment, my thirst was gone.

Denny Christian, director of technical support for the Center for Cold Ocean Resource Engineering (C-CORE) in St. John's, is an iceberg guy. Whenever I asked around on the subject of icebergs, his name came up. C-CORE is a company that handles extreme-conditions research-and-development problems, many having to do with oceanic ice. Denny Christian has photographed icebergs, towed them, measured them, and done experiments on them. When he was younger, he used to put crampons on his feet and climb on them. With other ice technicians, he has landed on them in helicopters and shut the engines down and walked around; he told me he would never do anything so reckless today. A few years ago, against his better judgment, he ferried out two scientists who wanted to climb on a berg and helped them aboard. Then he stood off in his skiff and watched. He heard a noise, saw some of the iceberg come out of the sea, and the next thing he knew, the scientists were clinging to the side of the ice they'd been standing on, 50 feet above the waterline. In another few seconds, the berg had gone over on top of them. By luck and quick boat-handling he was able to fish them out alive.

Denny Christian is tall, stoop-shouldered, snub-nosed, in his mid-sixties. His ancestors were Norwegian. When he strongly wishes to make a point, he widens his eyes with an attention-getting, Norwegian ferocity; other than that, his manner is laconic and mild. Usually he has the stoic patience you get from being around the unobligingness of nature. The only person he ever threw out of his office was a man who went on in great detail about a scheme he had to tow icebergs someplace or other with a submarine he planned to buy. Listening to crazy iceberg schemes is a recurring part of Denny Christian's job.

He and I sat in his office talking icebergs and going through folders of iceberg photos he has taken in his twenty years at C-CORE. The icebergs came in every category of shape and featured many natural parodies of architectural styles from caveman days to now. Some bergs suggested cliff dwellings, some castles; some were like fortresses, or space needles, or ultramodern Jetsons-type mansions in Beverly Hills. A striking iceberg that I had seen photos of before had two foothill eminences joined at the top by a soaring St. Louis Gateway Arch of ice. I stopped at a photo of a large tabular berg with men in red carrying chainsaws and clambering on it.

"These guys were some French environmentalists who wanted to carve a big iceberg in the shape of a whale with other endangered species riding on its back, and then send it down the East Coast of America to make a point about saving the environment," Denny said. "I helped them with the project, but I told 'em, 'You better wear these bright red life jackets so we can find you if you fall in and *die.*'" His eyes went wide. "Fortunately, the Gulf War came along about that time and their money ran out, and we never heard from them again."

Right now most of the work Denny does at C-CORE is for a group of oil companies that have drilling platforms in the North Atlantic about 350 miles east of St. John's. The oil drillers are worried about icebergs crashing into their platforms or sinking their tankers. Among other problems, C-CORE studies the likelihood of iceberg impacts, how damaging they might be, and how to avert them. Two summers ago Denny harnessed some small icebergs and smashed them into a steel panel rigged with instruments, to learn more about such collisions. Recently he has been studying the iceberg-tracking effectiveness of high-frequency radar that can see over the horizon out to about 250 miles. Every so often he charters a plane and flies out to check the radar's accuracy with his own eyes. When he told me he planned to take one of these "ground-truthing" flights in the next day or two, I asked if I could go along.

First we had to drive to Gander, about four hours from St. John's, to get the charter plane. Michelle Rose, a college intern at C-CORE who assists Denny sometimes, accompanied us. The wind was blowing hard when we reached the airport, but the charter pilots didn't seem to mind. Soon we were bumping and bouncing in a twin-engine plane through the updrafts over the North Atlantic, out of sight of land. Then the plane descended to a low altitude, the turbulence subsided, and we were cruising steadily 500 feet above the waves.

Denny had given the pilots GPS readings corresponding to places where the radar had indicated icebergs. The first one appeared on the plane's right side — a blocky berg with a line of dirt on it like a slash. Denny said that was glacial till that the glacier had scraped from the ground before calving. This next one, a pinnacle berg, had two bright blue lines on it that formed a V. Blue lines, he said, are from meltwater that flows into glacial crevices and freezes there; ice from meltwater lacks the bubbles in glacial ice and reflects blue. As we passed each berg, Denny photographed it and shouted data — shape, size, GPS coordinates — and Michelle Rose wrote them down on a clipboard. The distribution of icebergs did not seem to fit the radar picture very closely, but that slipped from everybody's mind.

Some icebergs we didn't get a good look at on the first pass, and so we swooped back for another. An iceberg like the Rock of Gibraltar caught the sun on a facet of its never-to-be-scaled white peak as we went by. A humpbacked berg nosed through the waves with homely persistence as if going somewhere. We checked out eight icebergs in all. Each was completely by itself in a sun-sparkled immensity of ocean, each had contours and particularities as distinctive as a face. The imitations they did of habitable places inspired a hopelessness that was almost sublime. Not much in nature makes you feel as lonesome as an iceberg does. Perhaps that's the reason we don't give them names: the uncomfortable parallel to our own diffusible selves.

Through Denny Christian I met a St. John's captain named Ed Kean. When the ocean still held lots of cod, four generations of Ed Kean's Newfoundland ancestors fished for them. Ed has an all-purpose 75-foot boat with twin diesel engines and a 5-ton crane amidships. The boat's name is *Mottak*, which he says is Inuit for "boat." He uses it for a miscellany of jobs. When a ship undergoing repairs offshore needs a large part ferried out to it, Ed does that. If a film crew shows up wanting to make a nature documentary for Japanese television, he takes them around. April through September, when icebergs are in local waters, Ed harvests iceberg ice. In the *Mottak* he searches for likely-looking pieces — the size of VW Beetles or small sports cars is about right — and when he finds them he sends his crewmen in a skiff and they wrap the iceberg pieces in nets and tow them back to the boat. Then Ed winches them out of the sea and into the hold. Back in port, he sells the ice to the Canadian Iceberg Vodka Corporation, which makes their triple-distilled Iceberg Vodka from it. Last year Ed sold the company almost a million liters of melted icebergs.

Before I knew what Ed looked like, I recognized him from his walk. He walks with the rolling gait of a sailor, feeling the floor for a moment with each foot before setting it down. Until I met him, I assumed that this nautical way of walking was extinct, or a conceit of literature. Ed roll-walked across the lobby of the Holiday Inn, where I was staying, and shook my hand. We'd arranged to meet there so I could go out on his boat and watch him gather iceberg ice. We drove out to where he docks his boat, in a little harbor northeast of St. John's. Fishermen seem to have an accent different from the crisp, clipped speech of other Newfoundlanders; as near as I can tell, it's part Irish, part Cockney. "Beautiful day for 'arvestin' oiceberg oice," Ed said, as he scanned the blue sky.

Also onboard the *Mottak* that day were two crew members, both named Tony, both tanned dark brown and with deeply lined, sunreddened eyes; Denny Christian, hoping to see an unusual iceberg Ed had promised to show him; Denny's wife, Thelma, an iceberg enthusiast like her husband; me; and a producer-director, a cameraman, a soundman, and an on-camera personality named Kevin, all from a cable television show called *The Thirsty Traveler.* This show began on a food network in Canada. In each half-hour episode it visits a different part of the world and samples an emblematic local alcoholic drink. In just the last few weeks, the crew had done sake in Japan, tequila in Mexico, and ouzo in Greece. Now they were doing Newfoundland and iceberg water vodka.

The ocean rose in gentle swells as the *Mottak* headed south along the coast, past an island called Bell Island. Denny joined me at the rail and said that the water we were on used to hide German submarines. Twice during World War II, a U-boat came into this narrow stretch between the island and the shore, waited for an opportunity, and torpedoed boats loaded with ore from the Bell Island iron mines. On clear days when the light is right, you can see the sunken tankers on the bottom. Scores of sailors died in the attacks, and each time the sub got away. Denny pointed to the coastline, indented all along its length with big and little coves. "People used to say that sometimes U-boats would pull into one of the coves, and then the German sailors would put on civilian clothing and come into St. John's and go to the movies," he confided.

After we'd been dieseling for a couple of hours, Ed pointed to an iceberg straight ahead. Another hour and a half passed as the berg grew larger with almost undetectable slowness. At about a quarter mile, you could see that the berg was immense, tabular, and completely brown on top; at a hundred yards, it looked as if badly financed developers had abandoned a housing project on it while still at the bulldozing stage. Dirt and gravel and large rocks were strewn all over in heaps and piles. Denny said he had never seen this much glacial till on an iceberg. The sun was hot by now, the wind had become brisk, and the melting edges sent mini-avalanches of gravel raining down. If an iceberg can be described as having a fit, this one was. Rocks clacked and splashed all around it, little waterfalls streamed from it every few feet, pieces of ice fell off the sides and shattered and bobbed. When iceberg ice melts quickly, the bubbles released from it make a sound like soda water fizzing. A piece of ice as big as a bedroom fell splintering into the sea, and its myriad fragments, when they came back to the surface, fizzed full-throatedly.

With a rifle, Ed fired a couple of shots at the iceberg. The idea was to knock loose a conveniently sized piece. From the white cliff, some ice chips flew, but nothing big fell down. After the second shot, Thelma shouted that she had seen a fox — that an arctic fox had come to the edge of the berg, peered over, and ducked back. Denny said it happens sometimes that foxes and other animals get stranded on icebergs when they drift free. All of us stood and waited for the fox to reappear at the ledge of dirty ice and gravel 30 feet above the boat deck, but it didn't. Thelma kept saying that she was sure she'd seen it, and that the fox had looked miserable and skinny.

After some to-ing and fro-ing at the request of the TV people — approaching the iceberg now this way, now that, for the best angle and light — Ed motored around to its shoreline side. Acres of ocean dotted with iceberg pieces extended in all directions. The Tonys launched the skiff and began to harvest; securing nets around unwieldy car-size pieces of ice floating in 41-degree water can't be as easy as they made it seem. As the cameraman filmed, Kevin, the on-air person, gamely jumped into the skiff and assisted with the net-handling. Overseeing from the deck, Ed supplied his own dialogue: "Oh, you'll 'ave to do better 'n that, boys!" he sang out when they brought a smaller-than-usual piece alongside.

The air filled with the pleasant click of grapples, the substantial thumps of boated ice, the whining of the crane, the slap of the waves. During a lull, an iceberg piece too big to bring aboard drifted up beside the *Mottak*'s stern. This piece was about the size of six parking spaces and almost level with the water. It had a section like a pulpit rising from one side. I lay facedown on the stern and leaned my head over to within inches of the living ice. At this distance it glistened with the dull, wet gray of cubes in an ice cube tray, and its surface was pitted with little depressions. With a push I could have slid onto it and drifted away. Just as I was thinking that, a corner of the stern knocked, not hard, against it. In an instant, the entire piece of ice split in two. The section with the pulpit angled precariously upward from the water and then rotated its former top to deep below. The movement had the succinctness of a wall revolving to reveal a hidden door. It sent a sympathetic, shivery rush from my spine to the backs of my knees.

Ed filled the hold with iceberg pieces and piled others on the deck. Late in the afternoon he headed for port, first making another close pass by the big iceberg. Thelma was watching out for the fox, and she and Denny and one of the Tonys caught a glimpse of it near where it had been before. They said it was running along the edge of the iceberg, trying to keep up with the progress of the boat. Thelma wanted to call the Humane Society and get them to send a helicopter to pick up the fox. Its prospects did not look good out there, miles from land and with no company but the gulls. Thelma stayed in the stern, scanning the iceberg with binoculars, as we motored away.

In a booklet written for iceberg enthusiasts, a St. John's engineer and history buff named Stephen E. Bruneau mentions a locally famous iceberg called the Virgin Berg. It was an iceberg "hundreds of feet high and bearing an undeniable likeness to the Blessed Virgin Mary" (as one account described it) that appeared off St. John's harbor in June of 1905. Thousands of people went to a high hill above St. John's to watch it drift by, and the fishing boats that came out to follow it became a flotilla as it continued on its way south along the coast. Catholic and Protestant Newfoundlanders took it as a sign — generally, as a mark of divine favor for the Catholic side.

Probably there were also a few skeptics who saw it as just a big piece of ice in the water. My own mild fanaticism for icebergs can sometimes be hard to explain to unconvinced acquaintances. I notice the bewilderment in their eyes, and it infects me: Why, exactly, should anyone get so worked up about a piece of ice? Like the Newfoundland faithful of a hundred years ago, each of us sees in an iceberg what we are disposed to see. And yet . . . if icebergs have no significance other than the fanciful notions we project, why do they look as they do? A white iceberg lit by the sun in a field of blue

ocean simply *looks* annunciatory. It might as well have those little
lines radiating from it — the ones cartoonists draw to show something shining with meaning. In its barefaced obviousness, an iceberg seems the broadest hint imaginable; but what is it a hint of?

When I went to the Ice Center, Judy Schaffier gave me a list of
ice-watch Web sites run by government agencies in a dozen countries. All across the United States and Canada are institutes and
university departments that devote some or all of their resources to
studying the world's ice. From the earth and the sky, ever more sophisticated instruments constantly record tiny changes in the ice. A
satellite that measures global sea levels, a key part of ice studies,
takes 500,000 sea level readings a day. Laser altimeters deployed
on satellites can detect the minutest shifts of ice position with essentially no error. And yet, with all this information flooding in,
broad conclusions are hard to come by. Science is specialization,
and almost no expert in a particular area wants to step out into
summary or generalizing. The experts tend to approach big ideas
like global warming with the greatest hesitancy. Apparently, nobody wants to be the one to tell us (for example) that our SUVs
have got to go.

What science will hazard instead of conclusions is a series of ifs.
Ten thousand or 20,000 square miles of ice broken off Antarctica
hardly diminishes that continent's total ice area of 5 million square
miles, but *if* the shedding of ice continues at an increasing rate,
and *if* the ice loss causes the glaciers inland to move faster toward
the sea, and *if* seawater flows in where glacial ice used to be and
reaches certain sub-sea-level parts of the continent and melts the
ice there, and *if* as a consequence the entire West Antarctic Ice
Sheet goes — well, then we would have the schnitzel, to speak
plainly. That much ice added to the ocean might raise world sea
levels anywhere from 13 to 20 feet. Such a rise would submerge
parts of the island of Manhattan and of the Florida peninsula, not
to mention many other coastal areas worldwide where about half
the planet's population lives.

News stories in the months following the *Titanic* disaster examined the preceding few years' weather conditions in the Arctic and
North Atlantic and concluded that a number of meteorological
features had been "unusual." The winter of 1910–11 had been "unusually" snowy and severe, the summer of '11 and the spring of '12

had been "unusually" warm, the icebergs had drifted "unusually" far south. Such descriptions bring to mind the TV weather forecasters who still speak of temperatures as being "unseasonably" warm; most of the years in the last two decades have been warmer than any recorded in decades previous, so what does "unseasonable" mean nowadays? In fact, from an expanded perspective of time, there is nothing unusual about so many icebergs being as far south as the place where the *Titanic* went down. Over the past millennia, as climatic events came and went, icebergs invaded the Atlantic in armadas. Stones dropped from melting Canadian icebergs have been found in sea sediments off the coast of Portugal. The North Atlantic climate over the last tens of thousands and hundreds of thousands of years has been characterized by periods of continent-wide glaciation, massive melting, and the intermittent huge discharge of icebergs.

As easily as The Iceberg fits a morality tale about human pride, it fits a climatologist's possible scenario of global warming ifs. If increased moisture held in the warmer atmosphere results in more precipitation in the North Atlantic region, and if that precipitation leads to more runoff into the ocean, and if warmer summers result in a greater melting of glacial ice in Greenland, and if lots more icebergs set sail, and if all that leads to a greatly increased amount of fresh water poured every year into the North Atlantic — then, possibly, the complicated process of tropical warm water rising, flowing north, giving off its heat, and cooling and sinking (the process that creates the great ocean currents) will be disrupted. And if those currents stop, and the heat they bring to northern Europe and parts of the United States and Canada no longer arrives, those regions will very likely become colder, like other places at similar latitudes that ocean currents don't warm.

Of course no one imagined any such scenarios back in 1912. Most people knew little or nothing about climatology. Yet it might turn out that a foreshadowing of major climate change in that part of the world was the real message The Iceberg carried for us out of the North Atlantic night.

And then again, maybe not. Discussions of global warming always deal in elaborate, scary possibilities, while always including enough disclaimers and unknowns to blunt the fear. Considering all that's at stake, I want to tell us what we should do immediately to change our lives and avert environmental catastrophe. But I can't bring myself to, somehow. All I can do is put in a good word for the sweeping conclusion and broad generalization. There aren't enough of them around — enough high-quality ones, I mean. I think we have yielded the sweeping-conclusion field to the wackier minds among us. Scenarios based on the Mothman Prophecies are colorful, but not a lot of help in the long run. Sensibly, most of us fort up behind our ever-growing heaps of information. But eventually, and maybe soon, we should draw a conclusion or two about where the globe is heading; and after that, maybe even act.

A lot of what is exciting about being alive can't be felt, because it's beyond the power of the senses. Just being on the planet, we are moving around the sun at 67,000 miles an hour; it would be great if somehow we could climb up to an impossible vantage point and just for a moment actually *feel* that speed. All this data we've got piling up is interesting, but short on thrills. Time, which we have only so much of, runs out on us, and as we get older we learn that anything and everything will go by. And since it will all go by anyway, why doesn't it all go right now, in a flash, and get it over with? For mysterious reasons, it doesn't, and the pace at which it proceeds instead reveals itself in icebergs. In the passing of the seconds, in the one-thing-after-another, I take comfort in icebergs. They are time solidified and time erased again. They pass by and vanish, quickly or slowly, regular inhabitants of a world we just happened to end up on. The glow that comes from them is the glow of more truth than we can stand.

JAMES GORMAN

Finding a Wild, Fearsome World Beneath Every Fallen Leaf

FROM *The New York Times*

CONCORD, MASS. — Dr. Edward O. Wilson, Pellegrino University Research Professor and honorary curator in entomology of the Museum of Comparative Zoology at Harvard University, winner of two Pulitzer Prizes and scientific honors too numerous to recount, is on his hands and knees, pawing in the leaf litter near Walden Pond. He eases into a half-sitting, half-reclining position and holds out a handful of humus and dirt. "This," he says, "is wilderness."

Just a dozen yards from the site of Thoreau's cabin, Dr. Wilson is delving into the ground with a sense of purpose and pleasure that would instantly make any ten-year-old join him. His smile suggests that at age seventy-three, with a troublesome right knee, he still finds the forest floor as much to his liking as a professor's desk.

These woods are not wild; indeed, they were not wild in Thoreau's day. Today, the beach and trails of Walden Pond State Reservation draw about 500,000 visitors a year. Few of them hunt ants, however. Underfoot and under the leaf litter there is a world as wild as it was before human beings came to this part of North America.

Dr. Wilson is playing guide to this microwilderness — full of ants, mites, millipedes, and springtails in a miniature forest of fungal threads and plant detritus in order to make a point about the value of little creatures and small spaces. If he wrote bumper stickers rather than books, his next might be "Save the Microfauna" or "Sweat the Small Stuff."

He begins his most recent book, *The Future of Life*, with a "Dear Henry" letter, talking to Thoreau about the state of the world and the Walden Pond woods. "Untrammeled nature exists in the dirt and rotting vegetation beneath our shoes," he writes. "The wilderness of ordinary vision may have vanished — wolf, puma and wolverine no longer exist in the tamed forests of Massachusetts. But another, even more ancient wilderness lives on."

In their world, centipedes are predators as fearsome as saber-toothed tigers, ants more numerous than the ungulates of African plains. And, in contrast to the vast preserves required by the world's most revered megafauna — grizzlies and elephants, jaguars and condors — maintaining biodiversity among the little creatures, shockingly rich in unexplored behavior and biochemistry, can be done on the cheap in relatively tiny patches, as small as a few acres, around the world.

Dr. Wilson is by no means turning from the grand plans for conservation. Indeed, he has suggested that 50 percent of the globe ought to be reserved for nonhuman nature. But he is a realist and, as he describes himself, "a lover of little things."

He has been turning over logs and rocks, looking into the world of insects and other tiny creatures, since he was a boy in Alabama and Florida. And he has not stopped. During the walk, he talks enthusiastically about a coming field trip to the Dominican Republic to investigate ants there, and about the publication this fall of a book-length monograph on the genus Pheidole, describing all 625 species of ants, including 341 new to science.

Researchers tend to share a kind of acquisitive passion to see, touch, grasp the world. Nothing passes without comment. As he strolled along the shore of Walden Pond on the way to the woods, Dr. Wilson spotted a butterfly and interrupted his discussion of the sizes of reserves needed for mammals, reptiles, and amphibians, complete with references and citations to scientific studies.

When the butterfly landed on the beach, Dr. Wilson stopped, leaning forward like a heron on the hunt, and peered at it. "It's hard to identify," he said. "It's a very beat-up little butterfly," probably the variety called a question mark, because of the design on its wings.

Having reached the woods and having begun to talk about what lay under the surface of the forest floor, he held the crumbled leaf litter and humus in his hand as if he were savoring what lovers of certain wines call the *goût de terroir*, or taste of the soil, a certain earthy specificity that the wine owes to the ground, not the grape. "When I go on a field trip," he said, "providing you can get me up to the edge of a natural environment, I usually don't go more than a hundred yards or so in, because when I settle down, immediately I start finding interesting stuff.

"This ground," he said, "we see it as two-dimensional because we're gigantic, like Godzilla. When you just go a few centimeters down, then you're in a three-dimensional world where the conditions change dramatically almost millimeter by millimeter. In one square foot of this litter you're looking at the tens of thousands of small creatures that you can still spot with your naked eye."

The ground was drier than usual, and Dr. Wilson speculated that the drought might have affected insects. Some he had hoped to see were not there. "If we looked long enough," he said, breaking open several rotting acorns, "we would find entire colonies of very small ants living in an acorn."

As he moved on, from log to log, he uncovered relatives of cave crickets, predatory rove beetles, termites, several varieties of ants, spiderlings, beetle larvae — not quite in the abundance he had hoped for. Still, the small wilderness was teeming.

"The exact perception of wilderness is a matter of scale," he writes in *The Future of Life*, going on to say that "microaesthetics" is "an unexplored wilderness to the creative mind." But he also notes that while microreserves are "infinitely better than nothing at all, they are no substitute for macro- and megareserves." He continues, "People can acquire an appreciation for savage carnivorous nematodes and shape-shifting rotifers in a drop of pond water, but they need life on the larger scale to which the human intellect and emotion most naturally respond."

Dr. Wilson is no sentimentalist, about nematodes or people. His proposals for microreserves are practical and hard-headed. The idea is fairly simple. While areas of nearly 25,000 acres are needed to have a good chance at preserving most large forms of life, plants and insects can sometimes be preserved in plots of 25 or even 2.5 acres.

In the Amazon, for instance, Dr. Wilson said, where the land is being savaged, "You'll see hanging on the side of a ravine some-

where a patch a farmer hasn't farmed, one hectare, to maybe ten."
Such a small area may not catch the eye of most conservationists,
he said, but, he added, "the entomologist and the botanist are
likely to say, hold on a minute."

A researcher, he said, may find species not found anywhere else,
and such plots can grow, with care and reseeding of the surrounding area. "You can do this in most parts of the world, in most developing countries, where a farmer or village elders would happily
take a thousand bucks for you to set aside ten or a hundred hectares and even hire them to help with the reseeding," he said.

But it is not just the developing world where biodiversity can be
preserved, bit by bit. City parks may hold small wonders. Even at
Walden Pond, in the midst of the Massachusetts suburbs, he said:
"Many of the species you find here are new to science. The basic biology of most of these things is poorly known or not known at all."

In the Walden woods live two relatively unknown ant species in
the genus *Myrmica*. "They've been noticed, but not named or described," he said. As to the nematodes and mites, he said, lifetimes
can be spent and careers can be made studying them.

It is not, of course, entomologists, or even weekend naturalists,
who need convincing about the richness of the forest floor. And
there are many reasons to try to preserve biological diversity at the
near-microscopic level and below. There is always potential economic value in new biochemical discoveries. There is real and present economic value in clean air and water, to which the plants and
insects and microbes contribute. In fact, Dr. Wilson points out, the
life of the planet is built on a foundation of tiny creatures. In ecological systems it is the giants that stand on the shoulders of mites.

Finally, there is the simplest argument of all, that life itself, in all
its variations, is astonishing and mysterious, and that humans have
a responsibility to preserve it.

Under one log turned over on our walk, the environment was
moist enough to provide a widely varied selection of insects. Dr.
Wilson picked up a gooey white worm without a trace of discom-
fort. "That is a fly larva, a maggot actually," he said. "I don't know
the kind of fly. It's very slimy."

Asked if he might be demonstrating the very reason many people do not like to turn over logs and dig under leaves, he laughed.
"I'll admit it's an acquired taste," he said. "Don't mistake me, I

don't expect legions of people, particularly Americans, going out and seeing how many different kinds of oribatid mites or fly larvae they can find.

"But," he continued, "they can get a feel, one way or the other, that what's at their feet is not dead leaves and dirt, but a living world with a diversity of creatures, some of which are so strange to the average experience that they beat most of the things you see in *Star Wars.*"

It is not so hard to imagine. Butterfly fanciers are legion. Dragonflies are now attracting watchers with binoculars. "But mainly," he said, "there's got to be some sense of the beauty and integrity and the extreme age of these areas."

At the next and last log, he pointed out a predatory rove beetle, a millipede (a detritovore), a spiderling, a nematode he caught only a glimpse of ("like a very tiny silvery strand"), more ants, and a wood cockroach. As he began to describe the thousands of inoffensive species of cockroach throughout the world, a siren went off on a nearby street, a reminder of where he was.

Dr. Wilson looked up from the timeless environment under the log. "That's sure a sound that Thoreau never heard."

CHARLES HIRSHBERG

My Mother, the Scientist

FROM *Popular Science*

IN 1966, Mrs. Weddle's first-grade class at Las Lomitas Elementary School got its first homework assignment: We were to find out what our fathers did for a living, then come back and tell the class. The next day, as my well-scrubbed classmates boasted about their fathers, I was nervous. For one thing, I was afraid of Mrs. Weddle: I realize now that she was probably harmless, but to a shy, elf-size, nervous little guy she looked like a monstrous, talking baked potato. On top of that, I had a surprise in store, and I wasn't sure how it would be received.

"My daddy is a scientist," I said, and Mrs. Weddle turned to write this information on the blackboard. Then I dropped the bomb: "And my mommy is a scientist!"

Twenty-five pairs of first-grade eyes drew a bead on me, wondering what the hell I was talking about. It was then that I began to understand how unusual my mother was.

Today, after more than four decades of geophysical research, my mother, Joan Feynman, is getting ready to retire as a senior scientist at NASA's Jet Propulsion Laboratory. She is probably best known for developing a statistical model to calculate the number of high-energy particles likely to hit a spacecraft over its lifetime and for her method of predicting sun spot cycles. Both are used by scientists worldwide. Beyond this, however, my mother's career illustrates the enormous change in how America regards what was, only a few decades ago, extremely rare: a scientist who's a woman and also a mother.

To become a scientist is hard enough. But to become one while running a gauntlet of lies, insults, mockeries, and disapproval — this was what my mother had to do. If such treatment is unthinkable (or, at least, unusual) today, it is largely because my mother and other female scientists of her generation proved equal to every obstacle thrown in their way.

My introduction to chemistry came in 1970, on a day when my mom was baking challah bread for the Jewish New Year. I was about ten, and though I felt cooking was unmanly for a guy who played shortstop for Village Host Pizza in the Menlo Park, California, Little League, she had persuaded me to help. When the bread was in the oven, she gave me a plastic pill bottle and a cork. She told me to sprinkle a little baking soda into the bottle, then a little vinegar, and cork the bottle as fast as I could. There followed a violent and completely unexpected pop as the cork flew off and walloped me in the forehead. Exploding food: I was ecstatic! "That's called a chemical reaction," she said, rubbing my shirt clean. "The vinegar is an acid and the soda is a base, and that's what happens when you mix the two."

After that, I never understood what other kids meant when they said that science was boring.

One of my mother's earliest memories is of standing in her crib at the age of about two, yanking on her eleven-year-old brother's hair. This brother, her only sibling, was none other than Richard Feynman, destined to become one of the greatest theoretical physicists of his generation: enfant terrible of the Manhattan Project, pioneer of quantum electrodynamics, father of nanotechnology, winner of the Nobel Prize, and so on. At the time, he was training his sister to solve simple math problems and rewarding each correct answer by letting her tug on his hair while he made faces. When he wasn't doing that, he was often seen wandering around Far Rockaway, New York, with a screwdriver in his pocket, repairing radios — at age eleven, mind you.

My mother worshiped her brother, and there was never any doubt about what he would become. By the time she was five, Richard had hired her for two cents a week to assist him in the electronics lab he'd built in his room. "My job was to throw certain switches on command," she recalls. "I had to climb up on a box to reach them. Also, sometimes I'd stick my finger in a spark gap for the edification of his friends." At night, when she called out for a glass of water, Riddy, as he was called, would demonstrate centrifugal force by whirling it around in the air so that the glass was upside down during part of the arc. "Until, one night," my mother recalls, "the glass slipped out of his hand and flew across the room."

Richard explained the miraculous fact that the family dog, the waffle iron, and Joan herself were all made out of atoms. He would run her hand over the corner of a picture frame, describe a right triangle, and make her repeat that the sum of the square of the sides was equal to the square of the hypotenuse. "I had no idea what it meant," she says, "but he recited it like a poem, so I loved to recite it too." One night, he roused her from her bed and led her outside, down the street, and onto a nearby golf course. He pointed out washes of magnificent light that were streaking across the sky. It was the aurora borealis. My mother had discovered her destiny.

That is when the trouble started. Her mother, Lucille Feynman, was a sophisticated and compassionate woman who had marched for women's suffrage in her youth. Nonetheless, when eight-year-old Joanie announced that she intended to be a scientist, Grandma explained that it was impossible. "Women can't do science," she said, "because their brains can't understand enough of it." My mother climbed into a living room chair and sobbed into the cushion. "I know she thought she was telling me the inescapable truth. But it was devastating for a little girl to be told that all of her dreams were impossible. And I've doubted my abilities ever since."

The fact that the greatest chemist of the age, Marie Curie, was a woman gave no comfort. "To me, Madame Curie was a mythological character," my mother says, "not a real person whom you could strive to emulate." It wasn't until her fourteenth birthday — March 31, 1942 — that her notion of becoming a scientist was revived. Richard presented her with a book called *Astronomy*. "It was a college textbook. I'd start reading it, get stuck, and then start over again. This went on for months, but I kept at it. When I reached page 407, I came across a graph that changed my life." My mother shuts her eyes and recites from memory: "'Relative strengths of the Mg_+ absorption line at 4,481 angstroms . . . from *Stellar Atmospheres* by Cecilia Payne.' Cecilia Payne! It was scientific proof that a woman was capable of writing a book that, in turn, was quoted in a text. The secret was out, you see."

*

My mother taught me about resonances when I was about twelve. We were on a camping trip and needed wood for a fire. My brother and sister and I looked everywhere, without luck. Mom spotted a dead branch up in a tree. She walked up to the trunk and gave it a shake. "Look closely," she told us, pointing up at the branches. "Each branch waves at a different frequency." We could see that she was right. So what? "Watch the dead branch," she went on. "If we shake the tree trunk in just the right rhythm, we can match its frequency and it'll drop off." Soon we were roasting marshmallows.

The catalogue of abuse to which my mother was subjected, beginning in 1944 when she entered Oberlin College, is too long and relentless to fully record. At Oberlin, her lab partner was ill prepared for the advanced-level physics course in which they were enrolled, so my mother did all the experiments herself. The partner took copious notes and received an A. My mother got a D. "He *understands* what he's doing," the lab instructor explained, "and you don't." In graduate school, a professor of solid state physics advised her to do her Ph.D. dissertation on cobwebs, because she would encounter them while cleaning. She did not take the advice; her thesis was titled "Absorption of Infrared Radiation in Crystals of Diamond-Type Lattice Structure." After graduation, she found that the "Situations Wanted" section of the *New York Times* was divided between Men and Women, and she could not place an ad among the men, the only place anyone needing a research scientist would bother to look.

At that time, even the dean of women at Columbia University argued that "sensible motherhood" was "the most useful and satisfying of the jobs that women can do." My mother tried to be a sensible mother and it damn near killed her. For three years she cooked, cleaned, and looked after my brother and me, two stubborn and voluble babies. One day in 1964 she found herself preparing to hurl the dish drain through the kitchen window and decided to get professional help. "I was incredibly lucky," she remembers, "to find a shrink who was enlightened enough to urge me to try to get a job. I didn't think anyone would hire me, but I did what he told me to do." She applied to Lamont-Doherty Observatory and, to her astonishment, received three offers. She chose to work part-time, studying the relationship between the solar wind and the magnetosphere. Soon she would be among the first to announce that the magnetosphere — the part of space in which

Earth's magnetic field dominates and the solar wind doesn't enter — was open-ended, with a tail on one side, rather than having a closed-teardrop shape, as had been widely believed. She was off and running.

My mother introduced me to physics when I was about fourteen. I was crazy about bluegrass music, and learned that Ralph Stanley was coming to town with his Clinch Mountain Boys. Although Mom did not share my taste for hillbilly music, she agreed to take me. The highlight turned out to be fiddler Curly Ray Cline's version of "Orange Blossom Special," a barn burner in which the fiddle imitates the sound of an approaching and departing train. My mother stood and danced a buck-and-wing and when, to my great relief, she sat down, she said, "Great tune, huh? It's based on the Doppler effect." This is not the sort of thing one expects to hear in reference to Curly Ray Cline's repertoire. Later, over onion rings at the Rockybilt Cafe, she explained: "When the train is coming, its sound is shifting to higher frequencies. And when the train is leaving, its sound is shifting to lower frequencies. That's called the Doppler shift. You can see the same thing when you look at a star: if the light source is moving toward you, it shifts toward blue; if it's moving away, it shifts toward red. Most stars shift toward red because the universe is expanding."

I cannot pretend that, as a boy, I liked everything about having a scientist for a mother. When I saw the likes of Mrs. Brady on TV, I sometimes wished I had what I thought of as a mom with an apron. And then, abruptly, I got one.

It was 1971 and my mother was working for NASA at Ames Research Center in California. She had just made an important discovery concerning the solar wind, which has two states, steady and transient. The latter consists of puffs of material, also known as coronal mass ejections, which, though long known about, were notoriously hard to find. My mother showed they could be recognized by the large amount of helium in the solar wind. Her career was flourishing. But the economy was in recession and NASA's budget was slashed. My mother was a housewife again. For months, as she looked for work, the severe depression that had haunted her years before began to return.

Mom had been taught to turn to the synagogue in times of trouble, and it seemed to make especially good sense in this case, because our synagogue had more scientists in it than most Ivy League universities. Our rabbi, a celebrated civil rights activist, was arranging networking parties for unemployed eggheads. But when my mother asked for an invitation to one of these affairs, he accused her of being selfish. "After all — there are *men* out of work just now."

"But Rabbi," she said, "it's my *life*."

I remember her coming home that night, stuffing food into the refrigerator, then pulling out the vacuum cleaner. She switched it on, pushed it back and forth across the floor a few times, then switched it off and burst into tears. In a moment, I was crying too and my mother was comforting me. We sat there a long time.

"I know you want me here," she told me. "But I can either be a part-time mama or a full-time madwoman."

A few months later, Mom was hired as a research scientist at the National Center for Atmospheric Research, and we moved to Boulder, Colorado. From then on, she decided to "follow research funding around the country, like Laplanders follow the reindeer herds." She followed it to Washington, D.C., to work for the National Science Foundation, then to the Boston College Department of Physics, and finally, in 1985, to JPL, where she's been ever since. Along the way, she unlocked some of the mysteries of the aurora. Using data from Explorer 33, she showed that auroras occur when the magnetic field of the solar wind interacts with the magnetic field of the Earth.

In 1974, she became an officer of her professional association, the American Geophysical Union, and spearheaded a committee to ensure that women in her field would be treated fairly. She was named one of JPL's elite senior scientists in 1999 and the following year was awarded NASA's Exceptional Scientific Achievement Medal.

Soon she'll retire, except that retirement as my mother the scientist envisions it means embarking on a new project: comparing recent changes in Earth's climate with historic ones. "It's a pretty important subject when you consider that even a small change in the solar output could conceivably turn Long Island into a skating rink — just like it was some 10,000 years ago."

The first thing I did when I came home from Mrs. Weddle's class that day in 1966 was to ask my mother what my father did. She told me that he was a scientist, and that she was a scientist too. I asked what a scientist was, and she handed me a spoon. "Drop it on the table," she said. I let it fall to the floor. "Why did it fall?" she asked. "Why didn't it float up to the ceiling?" It had never occurred to me that there was a "why" involved. "Because of gravity," she said. "A spoon will always fall, a hot-air balloon will always rise." I dropped the spoon again and again until she made me stop. I had no idea what gravity was, but the idea of "Why?" kept rattling around in my head. That's when I made the decision: the next day, in school, I wouldn't just tell them what my father did. I'd tell them about my mother too.

BRENDAN I. KOERNER

Embryo Police

FROM *Wired*

FOR ALAN AND LOUISE MASTERTON, the death of their daughter, Nicole, had a uniquely cruel twist. It was terrible enough that the three-year-old succumbed to burns suffered in an accident at the family's Monifieth, Scotland, home in 1999. But for the Mastertons, Nicole was more than just a cherished child — she was a chromosomal miracle. The couple had spent fifteen years trying to conceive a girl, bearing four sons in the process. When Nicole finally arrived in 1995, the Mastertons considered their prayers answered and their family complete. Louise had her tubes tied.

A month after Nicole's death, a heartbroken Alan began posting messages on Usenet's fertility groups. "We know that if we had another 100 children, none could replace Nicole," he wrote. "But she has left such a huge emptiness in all our hearts, we feel (possibly selfishly) desperate to have another little girl to love and cherish, to let her live the life that Nicole was so cruelly denied." The Mastertons wanted another go at parenthood, provided they could be guaranteed a female child. The note ended with an appeal for leads on in vitro fertilization clinics that could identify an embryo's sex prior to implantation in the womb.

Had the Mastertons lived in the United States, the Wild West of reproductive medicine, their quest would have sparked little outcry. Fertility clinics in the United States aren't subject to government licensing — unlike, say, tattoo parlors, veterinary hospitals, and, of course, individual doctors. Any family with $30,000 to spend can order a lab-concocted boy or girl.

The British, however, are more cautious about meddling in Nature's affairs. In the United Kingdom, a government panel known as the Human Fertilization and Embryology Authority tightly regulates the fertility industry. The HFEA, the world's first official overseer of reproductive technology, forbids human bioengineering for "social purposes," a catchall ban that largely disallows gender selection. If the Mastertons wished to proceed, they would need to change the HFEA's mind.

Currently, the HFEA is unique to Britain — but not for long. Canada's national health service, Health Canada, is considering a biotech jury based on the HFEA. Japan, Australia, and several European nations also have inquired about transplanting the authority's model. Ruth Deech, the HFEA's chair, spent last October in Japan, lecturing on the finer points of IVF regulation. There has even been a spark of interest in the freewheeling U.S.: James Childress, a bioethicist from the University of Virginia, lauded the HFEA at a September Senate hearing, noting that "Congress might consider that model for our society as well."

Much of the admiration stems from the HFEA's willingness to confront even the stickiest moral quandaries with a cold, rational eye. As Britain's de facto guardian of the human blueprint, the HFEA takes seriously its obligation to look past emotional appeals in the interest of the big picture. "We are sympathetic to the difficulties of these couples, of course," says Jane Denton, a former nurse and one of the authority's twenty-one members. "But we do have to look at the overall principle of whether or not it's appropriate to allow some of these techniques. Because if it goes ahead in one particular case, it is eventually going to become widespread."

Most of all, the HFEA fears the slippery slope, a long-debated Ethics 101 concept best summed up in this case as "Today sex-specific children, tomorrow a bioengineered master race." Perhaps allowing the Mastertons to manufacture a substitute daughter would be a first step toward a *Gattaca*-like future of made-to-order babies, scrubbed clean of diseases and endowed with sparkling blue eyes — a world in which eugenics is just another branch of science.

Alan Masterton was primed for the challenge of persuading the HFEA. A feisty Scot who bears more than a passing resemblance to the tough-guy actor Brian Dennehy, Masterton likes to point out his family's motto: *Ea quibus credimus defendimus, dum ceciderit ultimus,* or "We fight for what we believe in, until the last of us falls." In the past, the HFEA has approved sex selection, but only in rare cases involving genetic diseases that strike a particular gender. Families with a history of hemophilia, for example, are given permission to specify female, because only males are afflicted with the illness. To Masterton, making a similar exception for mental health reasons didn't seem like much of a stretch. "The void that Nicole has left in our home and in our hearts can never be filled," he says. "But another female child would help us all heal a little bit better."

A Dundee University law student at the time, Masterton spent three months compiling a thirty-page pamphlet containing his point-by-point argument in favor of sex selection. On the cover was a picture of a smiling, pigtailed Nicole and an inscription: "The joy and happiness she brought into our lives, her spirit, her place in our family and our hearts are the driving force behind this appeal." Masterton had twenty-one copies bound — one for each HFEA member — and sent them to London, hoping that the snapshot of a smiling Nicole would melt some hearts.

Unlike the United States, Britain was relatively well prepared for the genetics revolution. The U.K. was the site of the first test-tube baby, Louise Brown, in 1978, and ever since has been trying to divine the murky implications of bioengineering. In 1984, the government-appointed Warnock Committee released a prescient report on the long-term problems that might arise from reproductive technology, everything from snake oil treatments and dangerous implantation procedures to the ethics of designer kids. "And then that committee suggested setting up another committee, which tends to be a very British way of doing things," says Deech, the HFEA's chair since '94 and an Oxford University administrator. The bioethics panel was created in 1991 to license and inspect IVF clinics, with an eye toward preventing rogue experimentation. Fertility doctors who are discovered engaging in unapproved practices — implanting too many embryos in patients, skirting the rules on sex selection — risk losing their licenses. Those who dare soldier on without licenses risk prison terms.

The wielders of this extraordinary power are, for the most part, ordinary folks. The HFEA's charter requires that at least half of the panel's members come from outside the medical community. And neither the chair nor vice chair can have any connection to the fertility industry. Current ranks include a journalist, an Anglican bishop, an accountant, a social worker, and a retired customs official. "People know we're not just a bunch of mad, profit-seeking doctors running around in white coats," says Sara Nathan, an ex-BBC producer and a member since 1998. "The laypeople are there to make sure the doctors don't let science run ahead of what society wants."

Members are recruited via newspaper ads, and hundreds of applications flood the national Department of Health. A rigorous interview process follows, designed to weed out the merely curious from the passionate. "Sometimes we feel we need a child psychologist, or someone with experience in counseling, or an ethicist, or a philosopher, or a clergyman," says Deech, who assists in the selection process. "We also look for people who do not have any sort of emotional baggage — I think it would be a mistake to have a member who would be crusading for a particular point of view."

Those who are accepted receive three-year, renewable terms, along with a packed schedule of committee powwows and inspection tours. Though most members work full-time at other jobs, they are expected to dedicate at least one day per week to their HFEA duties. The panel meets monthly at the agency's headquarters near London's Liverpool Street.

Passions often flare around the huge oval table, as on this day, when members debate whether new fertility treatments merit the HFEA's seal of approval. The gatherings are closed to the public; only skeletal minutes are published, and individual comments are not recorded. Nathan admits "the meetings can get heated." She recalls her first one, which focused on egg sharing, an arrangement by which an infertile (usually wealthy) woman offers to pay for another woman's IVF procedure in exchange for a few spare ova. Skeptics at the table argued that such arrangements violated a fundamental principle of egg donation — that it should be altruistic — and could pave the way for a black market in gametes. "We spent a long time drawing up egg-sharing guidelines," says Nathan, "to protect women from overly enthusiastic medical staff."

In addition, members supervise inspection teams, which monitor the nation's seventy-four licensed fertility clinics. At least once a year, an HFEA squad will visit each clinic to see that records are up-to-date and gametes are safely stored. The inspectors also make sure the clinic isn't inflating its success rate to impress potential clients. Even under the best of circumstances, only 25 percent of infertility patients will ever give birth.

The HFEA's task was difficult enough during its early-'90s infancy, when the fertility industry was a rather simple affair. But as the number of British couples seeking treatment doubled over the decade to more than 30,000 annually, the panel's scope broadened. Few Britons had heard of the authority before 1995, the year a man named Stephen Blood contracted a fatal case of meningitis. Moments before his death, his wife, Diane, persuaded doctors to use a technique known as electro ejaculation to extract his sperm. She insisted the couple had talked of starting a family, and she hoped to create his children posthumously. But because Diane was unable to obtain explicit consent from her comatose husband, the HFEA tried to block her reproductive efforts, forbidding any U.K. clinic to impregnate her with the frozen sperm.

"Here was a man in a coma, at death's door," says Deech. "He didn't know that his sperm was being taken from him . . . If there is one fundamental principle in common law, it is that you never do anything to a person when they are unconscious and without consent." Diane Blood's plight generated a tremendous amount of public sympathy, and she contested the HFEA's decision according to a European treaty that permits freedom of movement for medical reasons. She was eventually allowed to take her husband's gametes to Belgium, where she was impregnated.

Deech views the Blood case as emblematic of the fertility industry's evolution and of the HFEA's as well. "When IVF started, I think it was imagined that we would be dealing with infertile couples who needed a baby to complete their families," says Deech. "But around about the early to mid-'90s, it began to move beyond that. Clinicians and patients could see the possibilities of extending fertility beyond menopause, of freezing sperm or embryos, posthumous babies, and so on. And it moved away from just treatment of infertility to, if you like, matters of convenience."

Now, with the human genome a more or less open book, Deech foresees the HFEA grappling with a new era. "We have now just moved on to a third phase, which is genetic engineering," she says. "Are these techniques going to be used for the improvement — or, some would say, the manipulation — of babies?"

Fertility technology is developing at a relentless pace. Ten years ago, an infertile male's odds of siring children were nil. Now, a procedure known as intra-cytoplasmic sperm injection, in which a single sperm is directly injected into an egg, gives 95 percent of those afflicted a shot at genetic fatherhood. With improved hormonal regimens, IVF enables women in their fifties, even their sixties, to bear children. A few mavericks are freezing the testicular or ovarian tissue of terminally ill prepubescents in the hope that technology might someday allow for the creation of sex cells.

The struggle to overcome infertility has brought researchers to the cusp of altering humanity's genetic heritage. As academics bicker over cloning, more efficacious technologies are quietly changing the rules of reproduction. One such method is cytoplasmic transfer, a process by which the damaged eggs of older women are repaired with injections of cytoplasm — the "jacket" that surrounds the egg — harvested from younger women. Since cytoplasm contains mitochondrial DNA, the resultant child inherits genetic material from several sources — the father, the mother, and the cytoplasm donor (or donors). Jacques Cohen, a researcher at the Institute for Reproductive Medicine and Science of Saint Barnabas, a New Jersey fertility clinic, has supervised the creation of fifteen such multiparent babies over the past three years. In the March 2001 issue of the journal *Human Reproduction,* he boasted that his experiment was "the first case of human germline genetic modification resulting in normal, healthy children." He failed to note that cytoplasmic transfer negates the most basic equation of mammalian reproduction — one male plus one female equals offspring.

Then there are the rodents. In March 2000, Japanese and American researchers announced that they had transplanted human ovaries into mice. Their goal was to "create an egg bank for patients suffering from infantile cancer who may survive into adulthood and want to have a child," explains project leader Akiyasu Mizukami. The breakthrough followed the 1999 claim of Nikolaos Sofikitis, a Greek doctor, who said he'd grown human sperm in rat testes.

Some bioethicists fear these pioneers are pushing humanity down that infamous slippery slope. One nightmare scenario was popularized by Princeton University molecular biologist Lee Silver in his 1998 book *Remaking Eden: How Genetic Engineering and Cloning Will Transform the American Family.* Silver predicted that wealthy families would have their offspring crafted in IVF clinics, where the tykes would be outfitted with genes that confer advantages in intelligence, health, and appearance. These "GenRich" would lord over the unendowed masses, the "Naturals," who would provide menial labor. "The GenRich class and the Natural class will become entirely separate species with no ability to crossbreed," Silver prophesied, "and with as much romantic interest in each other as a current human would have in a chimpanzee."

That split seems distant, but inklings of a remade Eden are beginning to appear. Last January, scientists at the Oregon Regional Primate Research Center reported that they'd succeeded in endowing a rhesus monkey, nicknamed ANDi, with jellyfish genes. They did so by injecting viruses laden with jellyfish DNA into the mother's unfertilized eggs. ANDi was a first, small step toward allowing doctors to imbue human eggs with genes that confer disease resistance, or even enhanced mental and physical traits.

The standard media reaction to ANDi's birth was horror — a *Saturday Night Live* sketch joshed that the discovery would "be reported in the *New England Journal of Evil*." But what loving parents would resist the chance to enhance their child's prospects for health and happiness? Lori Andrews, the author of *Future Perfect: Confronting Decisions About Genetics,* cites a Boston University Medical School study that found that 12 percent of women say they would abort a fetus with a genetic predisposition to obesity. And American IVF clinics are attracting well-heeled clients by selling gametes derived from Nobel laureates or Yale graduates, services that could conjure up a new tort. "If a woman gets sperm from a Nobel sperm bank," says Andrews, "and $E = mc^2$ isn't the first thing out of the child's mouth, will they sue?"

That legal question is typical of the Solomonic dilemmas wrought by the new technology. One of the HFEA's prime directives, for example, states that authority decisions must take into account both the welfare of the embryo and the welfare of existing children. But things get murky when those two obligations conflict, as is the situation with Raj and Shahana Hashmi, a couple whose case is being reviewed by the HFEA. Their two-year-old son, Zain, suffers from beta thalassemia major, a rare blood disorder that is invariably fatal. Zain endures a harsh treatment regimen of four blood transfusions and five marathon drug-infusion sessions per week. Yet his condition is worsening, and he'll soon die without a stem cell transplant.

The national registry contains no suitable donors, so the

Hashmis are considering a revolutionary IVF procedure to help their son. Using a pre-implantation genetic diagnosis, a doctor would select an embryo with compatible tissue from among Shahana's many fertilized eggs and then harvest the stem cells for Zain from the resulting child's umbilical cord. This seems to have worked for a Colorado couple, the Nashes, whose daughter, Molly, suffers from Fanconi anemia, a blood disease that leads to bone marrow failure. The Nashes used in vitro fertilization and PGD to select a tissue-matched son, Adam, who was born last October; an infusion of stem cells from her brother has given Molly a 90 percent chance of survival.

Despite the time-sensitive nature of the case, the HFEA is being judicious. Panelists consider the Hashmi decision fraught with ethical perils, most of them related to a frightful vision of babies rolling off a conveyor belt and being stamped SPARE PARTS. "If the first child needed a bone marrow transfusion, for example, or needed another organ, would compatibility mean the second child was always expected to provide these things?" asks the Right Reverend Michael Nazir-Ali, an HFEA member and the bishop of Rochester. One potential solution the HFEA has considered is making the tissue-matched offspring a ward of the court, thereby curtailing the parents' harvesting powers. A final decision is expected early this year, but in the meantime, the HFEA has denied the Hashmis' request to freeze eggs in anticipation of the ruling.

Though no polls have been taken, panel members claim that the majority of Britons support their go-slow approach. "What we do is certainly made easier by the fact that we have so much support from the general public," says Peter Mills, a spokesperson for the HFEA and one of thirty-four full-time staffers assigned to it. Clinicians, however, frequently rail against the panel, saying it's overly timid. Doctors howled when the authority held up the approval of intra-cytoplasmic sperm injection due to anecdotal evidence that the ensuing babies are prone to chromosomal abnormalities (likely inherited from their infertile fathers). The HFEA tarried despite the procedure's acceptance in the U.S. and Europe, which played host to hundreds of British "fertility tourists" throughout the 1990s. "The problem with the HFEA is that they have to be certain everything is safe before it is used," grumbles Simon Fishel, director of the Nottingham-based Centres for Assisted Reproduction.

"If everybody else was inhibited in the same way, no progress would be made. Indeed, some would argue that Louise Brown would never have come into existence with the current act."

The HFEA's influence over a controversial field of medicine, coupled with the secrecy of its deliberations, has made it a particularly inviting target in the press. Tom Utley, a popular *Daily Telegraph* columnist, wrote in November 1999: "It is hardly an exaggeration to say that, every time these 21 people meet, they are required to make decisions that in an earlier age were left to God."

The Fleet Street tabs were similarly unkind to the Mastertons, accusing the couple of reviving Nazi eugenics and treating kids like consumer items. "How long before every child born in Britain is as, one to another, a foil-wrapped pack of tomatoes?" wrote one *Daily Mail* scribe. Alan Masterton couldn't understand the fuss — he's heard of cases in which Britons discovered a fetus's gender via amniocentesis and then aborted. Why should Louise be denied the opportunity to select the sex of her child merely because she required IVF?

Alan made repeated requests — twenty-six by his count — to state his case in person. HFEA rules bar public participation at the panel's meetings, but chief executive Suzanne McCarthy assured him that his case would be considered. Finally, in January 2000, the issue came up at the authority's monthly conclave. The committee decided, in effect, not to hear the case at all. If the Mastertons wanted the HFEA to reconsider the use of PGD for sex selection, they would have to persuade a British clinic to file a licensing application on their behalf. "We sorted out the law," says Deech, "which was that they should really make their case to a clinic. And if a clinic really wants to espouse it, they should apply to us."

The decision left the Mastertons in a tight spot. They needed to find a clinic willing to challenge the HFEA's ruling. However, obtaining a license is an arduous process that few clinics would undertake without being fairly certain their application would be accepted — and, in this case, such an outcome was highly unlikely. Alan Masterton was enraged. "I was promised right up to the day before the meeting that my case would be seen by all twenty-one members of the committee," says Masterton, who accuses Deech of making the decision unilaterally. She denies this. With no formal appeals process, the family's options were limited. They requested a parliamentary investigation, and last May an ombudsman concluded that the HFEA did err in promising the Mastertons that an individual appeal would be heard. But the ombudsman has no power to force reconsideration.

The HFEA based its ruling on a technicality, but it's clear that many members harbored qualms about the Masterton case. Nazir-Ali believes the sex-selection boundary is best left uncrossed. "There's a fear that sex selection for social purposes will discriminate against women, in effect," he says. "There is also the fear that a child may be treated as a means to an end, rather than an end in itself, particularly if that child is seen as replacing a child who died in tragic circumstances. It's a heavy burden to bear if you're there as a replacement, not for your own sake."

Still desperate for a daughter, the Mastertons joined the hordes of British fertility tourists who go elsewhere for treatments deemed too risky or offensive back home. The richest travel to the United States; those of lesser means head to Italy, also an unregulated frontier of reproductive medicine. In July of 2000, with £6,000 borrowed from friends, the Mastertons visited Biogenesi, a Roman clinic, where doctors removed three eggs from Louise's ovaries. Two of the eggs were immature, so they couldn't be used — a common complication for a woman in her forties. The surviving ovum was successfully fertilized with Alan's sperm and then screened to ensure that the telltale Y chromosome was absent.

No dice. It was a male. And it was the last straw.

Lacking funds for another treatment cycle, the Mastertons donated the embryo to an infertile couple. And then they headed home, exhausted and broke. They've retained a London lawyer to research whether the family can file an appeal in accordance with Britain's Human Rights Act, which guarantees a fair judicial review. Alan knows it's a long shot, and his natural optimism has been supplanted with bitterness. "These people are coldhearted, uncaring bastards," he says, his Scottish brogue thickening with anger. "They look after their own asses and care not a jot for anyone else. Because of the shabby way my family has been treated by these deceitful people, I will crusade against the way this organization operates until it is changed to be more people-friendly and accountable."

The HFEA's handling of the Masterton case still riles Simon Fishel too, who favors a regulatory body staffed by more seasoned scientists in addition to laypersons. "If you and your medical practitioner want to do some chemical jiggery-pokery for you to have the gift of your own child, who else should have a say in that decision?" he asks. "Does it matter that your next-door neighbor has an opinion on how you reproduce?"

Absolutely, says HFEA member Sara Nathan. "Fertility doctors are not just curing somebody's arm ache — they're creating new people," she says. "And those people are going to live in society, and they'll have mutual responsibilities with society. That's why we need regulation that is both scientific and ethical."

The next big test for that credo will likely be the use of PGD for science beyond sex selection. By analyzing a sliver of embryonic cells, doctors can now determine whether an embryo contains genetic flaws and then refrain from implanting it in the womb. Of the 10,000 or so genetic disorders that can afflict an embryo, only a few are currently subject to tests — primarily grave conditions such as cystic fibrosis or Duchenne's muscular dystrophy. But if the evolution of pre-implantation genetic diagnosis mirrors that of other reproductive technologies, doctors could soon be able to identify whether a child is predisposed to middle-aged cancer, late-onset deafness, or even premature baldness.

Nevertheless, if a test were developed that could detect a propensity for severe late-onset diseases, the ever-cautious HFEA would be disinclined to approve its use. "By the time a person is forty-five, there might be a cure for heart disease, or that person may be a Beethoven who's written I don't know how many symphonies, or married and had beautiful children who are free of heart disease," says Deech. "People are fearful of the search for perfection, and in the public's mind there is a broad distinction between avoidance of a very serious disease that a parent would dread, and trying to choose an embryo who will be perfect."

That philosophy baffles Simon Fishel. "You either have to be on the side of medicine and scientific development or on the side of natural selection. You cannot just say, well, with regard to appendicitis, it's OK to intervene, but in regard to human reproduction, we should just leave it up to divine providence." If bioengineering can spare a human the burden of, say, diabetes, Fishel cannot imagine a good reason for not taking action. And if this science can also be used to select a baby's gender, then so be it.

But what about sparing a toddler the dangers of asthma? Or an adult the horrors of clinical depression? The thorniness of these questions has opened a debate in the United States about whether to let fertility doctors act as sole arbiters of what's best for humanity. There are federal agencies that regulate airwaves, power plants, highways, even harbor buoys. Why not a biotech jury to safeguard life itself? It all seems so sensible.

Not to everyone, though. "I am aware that your government has been discussing possibly following the HFEA model," says Alan Masterton, turning a bit gruff. "My heartfelt advice, born of experience, is don't do it."

ELIZABETH KOLBERT

Ice Memory

FROM *The New Yorker*

ICE, LIKE WATER, flows, and so the North Greenland Ice-core
Project, or North GRIP, lies in the center of the island, along a line
known as the ice divide. This is a desolate spot eight degrees north
of the Arctic Circle, but, thanks to the New York Air National
Guard, not actually all that difficult to reach. The research station
is open from mid-May to mid-August, and every season the Guard
provides some half-dozen flights to it, using specially ski-equipped
LC-130s. The planes, also outfitted with small rockets, can land directly on the ice, which stretches for hundreds of miles in every direction. (To the extent that there is a military justification for the
flights, it is to keep pilots in practice; however, the main purpose
of practicing seems to be to make the flights — an arrangement
whose logic I could never quite fathom.) This past June, I flew up
to North GRIP on a plane that was carrying several thousand feet
of drilling cable, a group of glaciologists, and Denmark's then minister of research, a stout, red-haired woman named Birthe Weiss.
Like the rest of us, the minister had to sit in the hold, wearing mili-
tary-issue earplugs.

One of the station's field directors, J. P. Steffensen, greeted us
when we disembarked. We were dressed in huge insulated boots
and heavy snow gear. Steffensen had on a pair of old sneakers, a
filthy parka that was flapping open, and no gloves. Tiny icicles
hung from his beard. First, he delivered a short lecture on the dangers of dehydration: "It sounds like a complete contradiction in
terms — you're standing on three thousand meters of water but it's
extremely dry, so make sure that you have to go and pee." Then he

briefed us on camp protocol. North GRIP has two computerized toilets, from Sweden, but men were kindly requested to relieve themselves out on the ice, at a spot designated by a little red flag.

Steffensen, a Dane, runs North GRIP along with his wife, Dorthe Dahl-Jensen, whom he met on an earlier Greenland expedition. Together with a few dozen fellow scientists — mostly Danes, but also Icelanders, Swedes, Germans, and Swiss, among others — they have spent the past six summers drilling a 5-inch-wide hole from the top of the ice sheet down to the bedrock, 10,000 feet below. Their reason for wanting to do this is an interest in ancient climates. My reason for wanting to watch is perhaps best described as an interest — partly lurid, but also partly pragmatic — in apocalypse.

Over the past decade or so, there has been a shift — inevitably labeled a "paradigm shift" — in the way scientists regard the Earth's climate. The new view goes under the catchphrase "abrupt climate change," although it might more evocatively be called neocatastrophism, after the old, biblically inspired theories of flood and disaster. Behind it lies no particular theoretical insight — scientists have, in fact, been hard-pressed to come up with a theory to make sense of it — but it is supported by overwhelming empirical evidence, much of it gathered in Greenland. The Greenland ice cores have shown that it is a mistake to regard our own, relatively benign experience of the climate as the norm. By now, the adherents of neocatastrophism include virtually every climatologist of any standing.

Abrupt climate changes occurred long before there was human technology and therefore have nothing directly to do with what we refer to as global warming. Yet the discovery that for most of the past hundred thousand years the Earth's climate has been in flux, changing not gradually or even incrementally, but violently and without warning, can't help but cast the global warming debate in new terms. It is still possible to imagine that the Earth will slowly heat up, and that the landscape and the weather will gradually evolve in response. But it is also possible that the change will come, as it has in the past, in the form of something much worse.

Greenland, the world's largest island, is nearly four times the size of France — 840,000 square miles — and except for its southern tip lies above the Arctic Circle. The first Europeans to make a stab at settling it were the Norse, under the leadership of Erik the Red, who, perhaps deliberately, gave the island its misleading name. In the year 985, he arrived with twenty-five ships and nearly seven hundred followers. (Erik had left Norway when his father was exiled for killing a man, and then was himself exiled from Iceland for killing several more.) The Norse established two settlements, the Eastern Settlement, which was actually in the south, and the Western Settlement, which was to the north of that. For roughly four hundred years, they managed to scrape by, hunting, raising livestock, and making occasional logging expeditions to the coast of Canada. But then something went wrong. The last written record of them is an Icelandic affidavit regarding the marriage of Thorstein Ólafsson and Sigrídur Björnsdóttir, which took place in the Eastern Settlement on the "Second Sunday after the Mass of the Cross" in the autumn of 1408.

These days, the island has 56,000 inhabitants, most of them Inuit, and almost a quarter live in the capital, Nuuk, about 400 miles up the western coast. Since the late 1970s, Greenland has enjoyed a measure of home rule, but the Danes, who consider the island a province, still spend the equivalent of $340 million a year to support it. The result is a thin and not entirely convincing First World veneer. Greenland has almost no agriculture or industry or, for that matter, roads. Following Inuit tradition, private ownership of land is not allowed, although it is possible to buy a house, an expensive proposition in a place where even the sewage pipes have to be insulated.

More than 80 percent of Greenland is covered by ice. Locked into this enormous glacier is 8 percent of the Earth's freshwater supply: enough, were it to melt, to raise sea levels around the world by more than 20 feet. Except for researchers in the summer, no one lives on the ice, or even ventures out onto it very often. (The edges are riddled with crevasses large enough to swallow a dog sled or, should the occasion arise, a five-ton truck.)

Like all glaciers, the Greenland ice sheet is made up entirely of accumulated snow. The most recent layers are thick and airy, while the older layers are thin and dense, which means that to drill down through the ice is to descend backward in time, at first gradually and then much more rapidly. A hundred and thirty-eight feet down, there is snow dating from the American Civil War; some 2,500 feet down, snow from the days of Plato; and, 5,350 feet down, from the time when prehistoric painters were decorating the caves of Lascaux. At the very bottom, there is snow that fell on Greenland before the last ice age, which began more than a hundred thousand years ago.

As the snow is compressed, its crystal structure changes to ice. (Two thousand feet down, there is so much pressure on the ice that a sample drawn to the surface will, if mishandled, fracture, and in some cases even explode.) But in most other respects the snow remains unchanged, a relic of the climate that first formed it. In the Greenland ice there is volcanic ash from Krakatau, lead pollution from ancient Roman smelters, and dust blown in from Mongolia on ice age winds. Every layer also contains tiny bubbles of trapped air, each of them a sample of a past atmosphere.

All across the Earth, there are, of course, traces of climate history — buried in lake sediments, deposited in ancient beetle casings, piled up on the floor of the oceans. The distinguishing feature of the Greenland ice, and what separates it from other ice, including ice extracted from the Antarctic, is its extraordinary resolution.

Even in summer, when the sun never sets, the snow doesn't melt in central Greenland, though during a clear day some of the top layer will evaporate. Then at night — or what passes for night — this moisture will refreeze. The immediate effect is lovely to behold: one morning, I was wandering around North GRIP at about five o'clock, and I saw the hoarfrost growing in lacy patterns underfoot. As the summer snow gets buried under winter snow, it maintains its distinctive appearance; in a snow pit, summer layers show up as both coarser and airier than winter ones. It turns out that even after thousands of years the difference between summer snow and winter snow can be distinguished. Thus, simply by counting backward it is possible to date each layer of ice and also the climatological information embedded in it.

The North Greenland Ice-core Project consists of six cherry-red tents arrayed around a black geodesic dome that was purchased, by mail order, from Minnesota. In front of the dome, someone has planted the standard jokey symbol of isolation, a milepost that shows Kangerlussuaq, the nearest town, to be 900 kilometers away.

Nearby stands the standard jokey symbol of the cold, a plywood palm tree. The view on all sides is exactly the same: an utterly flat stretch of white which could be described as sublime or, alternatively, as merely bleak.

Beneath the camp, an 80-foot-long tunnel leads down to what is known as the drilling room. This chamber has been hollowed out of the ice, and inside the temperature never rises above 14 degrees. A few years ago, foot-thick pine beams were added to reinforce the ceiling, but the weight of the snow piling up on top has grown so great that the beams have splintered. Because of the way the ice moves, the chamber, which is lit by overhead lights and filled with electronic equipment, is not just being buried but is also slowly shrinking and at the same time sinking.

Drilling begins at North GRIP every morning at eight. The first task of the day is to lower the drill, a 12-foot-long tube with big metal teeth on one end, down to the bottom of the borehole. Once in position, the drill can be set spinning so that an ice cylinder gradually forms within it. This, in turn, can be pulled up to the surface by means of a steel cable.

The first time I went down to watch the process, a glaciologist from Iceland and another from Germany were manning the controls. At the depth they had reached — 9,680 feet — it took an hour for the drill just to descend. During that period, there was not much for the two men to do except monitor the computer, which sat on a little heating pad, and listen to Abba. The ice near the bottom of the hole was warmer than expected, and it had been breaking badly. "The word 'stuck' is not in our vocabulary," the Icelander, Thorsteinn Thorsteinsson, told me with a nervous giggle. Eventually, the drillers managed to pull out a short piece of core — about 2 feet — to show Birthe Weiss, who arrived in the chamber wearing a red snowsuit. To me, it looked a lot like a 2-foot-long cylinder of ordinary ice, except that it was heavily scored around the edges. It was made up of snow that had fallen 105,000 years ago, Thorsteinsson said. Weiss exclaimed something in Danish and seemed suitably impressed.

After a piece of core comes up, it is packed in a plastic tube, put in an insulated crate, and shipped out on the next LC-130 to Kangerlussuaq. From there it is flown to Denmark, where it is stored in a refrigerated vault at the University of Copenhagen, to be cut up later for analysis. Inevitably, more researchers want a piece of the core than there are pieces to give out. A small library of papers has been written on the various gases and dust particles and radioactive byproducts that have been trapped in the ice. These papers have shown that the concentration of greenhouse gases in the atmosphere has fluctuated over time, and that these fluctuations have occurred roughly in tandem with changes in the climate. But the crucial insight has to do with the ice itself.

Water occurs naturally in several isotopic forms, depending on the hydrogen and oxygen atoms that joined together to make it. Typically, hydrogen has one proton and an atomic weight of 1, but when its nucleus also contains a neutron and it has an atomic weight of 2, it produces heavy water, which is used in nuclear reactors. Oxygen generally has eight neutrons and an atomic weight of 16, but it also comes in another stable version, with two extra neutrons and an atomic weight of 18. In any given water sample, the lighter ^{16}O atoms will vastly outnumber the heavier ^{18}O atoms, but by how much, exactly, is variable. In the early 1960s, Willi Dansgaard, a Danish chemist, proved that the ratio between the two in rainwater was related to the temperature. Dansgaard took samples of rain from around the world and demonstrated that, by running them through a mass spectrometer, he could in most cases arrive at the average temperature of the spot where they had fallen. Subsequently, he showed that this same technique could be applied to ice and, in particular, to the Greenland ice sheet.

Going back over the past 10,000 years, the Greenland ^{18}O record shows lots of bumps and squiggles. There is, for example, a slight but perceptible increase in temperature in the early years of the Middle Ages, which leads to what has become known as the Medieval Warm Period, when the English planted vineyards and the Norse established their Greenland settlements. And there's a dip some 600 or 700 years later, corresponding to the Little Ice Age, which killed off the vineyards and, most likely, led to the demise of the Greenland Norse. But the variation is limited. Between the Medieval Warm Period and the Little Ice Age, Greenland's average temperature fell by only a few degrees. Its average temperature today, meanwhile, is not very different from what it was 10 millennia ago, when our ancestors stopped doing whatever it was that they had been doing and learned to plant crops.

It's hard to look much farther back in the record, however, without feeling a little queasy. About 20,000 years ago, the Earth was still in the grip of the last ice age. During this period, called the Wisconsin by American scientists, ice sheets covered nearly a third of the world's landmass, reaching as far south as New York City.

The transition out of the Wisconsin is preserved in great detail in the Greenland ice. What the record shows is that it was a period of intense instability. The temperature did not rise slowly or even steadily; instead, the climate flipped several times from temperate conditions back into those of an ice age and then back again. Around 15,000 years ago, Greenland abruptly warmed by 16 degrees in fifty years or less. In one particularly traumatic episode some 12,000 years ago, the mean temperature in Greenland shot up by 15 degrees in a single decade.

If we go back farther still, the picture is no more comforting. Even as much of Europe and North America lay buried under glaciers, the temperature in Greenland was oscillating wildly, sometimes in spikes of 10 degrees, sometimes in spikes of 20. In an effort to convey the erratic nature of these changes, Richard Alley, a geophysicist who is leading a National Academy of Sciences panel on abrupt climate change, has compared the climate to a light switch being toyed with by an impish three-year-old. (The panel recently issued a report warning of the possibility of "large, abrupt, and unwelcome" climate changes.) He has also likened it to a freakish carnival ride. "Dozens of rapid changes litter the record of the last hundred thousand years," he observed. "If you can possibly imagine the spectacle of some really stupid person (or, better, a mannequin) bungee jumping off the side of a moving roller-coaster car, you can begin to picture the climate."

The first Greenland ice core was drilled in the mid-1960s at a U.S. military installation called Camp Century. The goal was not to challenge established views of the Earth's climate. Rather, the core was an instance of what Thomas Kuhn, in his famous essay "The Structure of Scientific Revolutions," called "normal science" — although it would perhaps be unfair to label anything associated with Camp Century as normal.

Built in 1959 in the northwestern corner of Greenland, the camp was a semisecret research station for a very cold war. It featured a tunnel 1,100 feet long and 26 feet wide, called Main Street, which led to dormitories, a ten-bed hospital, a mess hall, a skating rink, and a store that sold perfume to send back home — all under the ice. (A favorite camp joke was that there was a girl behind every tree.) Powering the enterprise was a portable nuclear reactor. "In an era in which it has become fashionable to describe the democratic countries as soft or lazy, the fantastic ice city is a wholesome answer to such nonsensical clichés," one particularly patriotic visitor reported. (The camp, which closed after a decade in operation, has since been obliterated by the movement of the ice.)

The U.S. Army Corps of Engineers led the camp's ice-coring effort. The Americans managed to drill their way right down to the bottom of the ice sheet, but when they were finished they didn't quite know what to do with the core they had produced. It fell to Chester Langway, a glaciologist who was working for the Corps's Cold Regions Research and Engineering Laboratory, to figure something out. Langway is now semiretired and operates a small antiques store on Cape Cod. He recalled traveling all around the country, attempting to drum up interest. "Some people looked at it and they said, 'That's just ice,' which it's not," he told me. Eventually, he and Dansgaard got in touch, and together the two men made the first study of the core.

At the time, one of the central questions in climate research was how ice ages began and how they ended. One theory, first worked out in detail in the 1920s by a Serbian astrophysicist named Milutin Milankovitch, was that glaciers advance and retreat in response to slight, periodic changes in the Earth's orbit. These changes alter the distribution of sunlight at various latitudes during various seasons, and Milankovitch predicted that the strongest effects would be observed at intervals of 19,000, 23,000, 41,000, and 100,000 years.

Dansgaard and Langway's study of the Camp Century core confirmed these so-called Milankovitch cycles but also gave evidence of the climate's carnival-ride-like reversals. This evidence was dismissed by many as an idiosyncrasy of the polar ice. Sigfus Johnsen was a student of Dansgaard's who worked with him on the Camp Century core, and he happened to be traveling to North GRIP at the same time I was. Johnsen is now sixty-one, with wispy white hair and pale blue eyes, and looks like a slightly dissolute Santa. He told me that the scientists working on the Camp Century core weren't sure themselves what to make of what they had found. "It was too incredible, something we didn't expect at all," he said.

It took fifteen years for anyone to drill another Greenland core. Largely because of Dansgaard and Langway's friendship, this second core was a European-American collaboration. It was drilled at an American radar base, Dye 3, a spot chosen for budgetary rather than scientific reasons, and from the outset everyone involved in the project knew that the location was a problem. Dye 3 was so close to the coast that the oldest layers of ice had mostly flowed out to sea. Nevertheless, the core confirmed all the most significant Camp Century results, demonstrating that findings which had seemed anomalous were at least reproducible. When the Dye 3 results were published in the early 1980s, they set off what is perhaps best described as an ice rush, and the spirit of international cooperation quickly broke down. The Europeans decided to drill a new core where the ice is most stable, along the ice divide, and the Americans decided to do the same thing some 20 miles away.

Theorists are still struggling to catch up with the data from those two cores, the first of which was completed in 1992 and the second a year later. No known external force, or even any that has been hypothesized, seems capable of yanking the temperature back and forth as violently, and as often, as these cores have shown to be the case. Somehow, the climate system — through some vast and terrible feedback loop — must, it is now assumed, be capable of generating its own instabilities. The most popular hypothesis is that the oceans are responsible. Currents like the Gulf Stream transfer heat in huge quantities from the tropics toward the poles, and if this circulation pattern could somehow be shut off — by, say, a sudden influx of fresh water — it would have a swift and dramatic impact. Computer modelers have tried to reproduce such a shutdown, with some success. But once the ocean circulation comes to a halt, modelers have had a hard time getting it to start up again. "We are in a state now where the more we know, the more it becomes clear how little we really understand about the system," the oceanographer Jochem Marotzke told me.

For at least half a million years, and probably a lot longer, warm periods and ice ages have alternated according to a fairly regular, if punishing, pattern: 10,000 years of warmth followed by 90,000 years of cold. The current warm period, the Holocene, is now

10,000 years old, and, all things being equal — which is to say had we not interfered with the pattern by burning fossil fuels — we should now be heading toward another ice age.

As a continuous temperature record, the Greenland ice gives out at about 115,000 years ago, at a moment in the climate cycle roughly analogous to our own — the end of the last interglacial period, which the Europeans call the Eemian and the Americans the Sangamon. What this part of the record suggests is disputed. The European core seemed to indicate that the period ended with a cataclysm even worse than the wild temperature swings that occurred at the end of the Wisconsin. During this cold snap, temperatures appear to have plunged from warmer than they are today to the coldest levels of the ice age, all within a matter of a few decades, and then to have climbed back up again, equally dramatically, a century or so later. The Europeans euphemistically dubbed this instant Eemian ice age Event One.

The Americans, however, determined that at the bottom of their core the ice had folded in on itself as it flowed, making accurate interpretation impossible. The two groups got together for a conference in Wolfeboro, New Hampshire, in 1995 and agreed on virtually everything except for Event One. Steffensen, North GRIP's field leader, recalls that the Europeans, who had rushed to publish their results, were crestfallen to have them discredited. He remembers going down to the hotel bar in Wolfeboro with some colleagues and thinking, The Eemian is dead; we have to bury it. Then, he told me, "we had a few drinks, and it came back to life."

The driving purpose behind North GRIP is to drill a core that will finally provide a clean record of the last interglacial period and validate Event One. Doing so would obviously be significant for several reasons: retrospectively, it would show temperature instabilities in yet another part of the climate cycle, and prospectively, it would seem to suggest a cataclysm in our own not too distant future. "In the first place, if we find this, it will scare the hell out of us," Sigfus Johnsen told me. But getting back to the Eemian remains a daunting technical challenge. In the summer of 1997, after the drillers at North GRIP had reached nearly a mile deep, the drill got stuck and could not be retrieved. The next summer they had to start all over again. They were almost 2 miles down when, in July 2000, the drill got stuck again. At that point, they poured antifreeze down the hole and eventually managed to yank the drill back up, but by then the weather was turning, and they had to close the station for the season. When I arrived, last June, they had finally finished bailing out the antifreeze and resumed drilling. Still, things were not going well. Something — presumably geothermal heat from some previously undetected "hot spot" — was warming the ice from below, making drilling extremely difficult.

Everyone I met at North GRIP was quite open about saying he believed — and hoped — that Event One had indeed taken place. Apparently, the prospect of having spent six summers up on the ice with essentially nothing to show for it was more disturbing than any fate that might lie in store for the planet. I found myself feeling torn as well. On the one hand, Event One did not sound like much to look forward to. On the other, it did seem to offer a certain perverse consolation: global warming versus Event One — either way, things were bound to end badly. I proposed this idea to Steffensen. Unimpressed, he pointed out that, if you believed the climate to be inherently unstable, the last thing you'd want to do is conduct a vast unsupervised experiment on it, and he went on to explain that it would be wrong to think of global warming and Event One as alternatives. It is entirely possible, if apparently paradoxical, that global warming could produce a precipitous cooling, at least in Europe and parts of North America, by, say, shutting down the Gulf Stream. It is also possible that it could push the climate into an unstable mode, leading, especially in the upper latitudes, to a period of wild temperature swings of the sort that characterized the end of the last ice age. Finally, it is possible that we have changed the atmosphere so much — carbon dioxide levels are approaching those of the age of the dinosaurs — that we will enter a new climate phase altogether. During the Cretaceous period, there were no major ice sheets, or ice ages, and much of the planet was covered with steamy swamps. To the extent that the historical record is any guide, the result of any climate change is unlikely to be a happy one. Steffensen recited to me an old Danish saying, whose pertinence I didn't entirely understand but which nevertheless stuck with me. He translated it as "Pissing in your pants will only keep you warm for so long."

Life at North GRIP, if not exactly comfortable, is at least well supplied. Lunch the day I arrived was a fish stew prepared in a delicate tomato base. In the midafternoon, there was coffee and cake; then, in the evening, cocktails, which were served in a chamber hollowed out of the snowpack, to relieve pressure on the drilling room. The German driller had provided a recipe for *Glühwein,* and everyone — scientists, graduate students, the Danish minister and her entourage, and the crew from the Air National Guard — was standing around in the dark, in cold-weather gear, drinking. ("Why do all the Danes I meet seem to come from Copenhagen?" I heard one of the pilots ask a young glaciologist.) For dinner, although I wasn't really hungry, I had a lamb chop in cream sauce, topped with diced leeks. At around midnight, the drillers finally emerged from under the ice. It was broad daylight outside, and inside the geodesic dome there was still a crowd drinking beer and smoking cigars.

As with so many recent discoveries about natural history, what seems, in the end, most surprising about the Greenland cores is exactly what might have seemed, at the outset, not to require any explanation at all. How is it that we happen to live in this, climatologically speaking, best of all possible times? On statistical grounds, it certainly seems improbable that the only period in the climate record as stable as our own *is* our own. And it seems, if anything, even more improbable that climatologists should make the discovery that we are living in this period of exceptional stability at the very moment when, by their own calculations, it is likely nearing an end.

But to approach the problem in this way is to fail to realize the extent to which we are ourselves a product of the climate record. Scientists were once puzzled by the evidence in lake sediments of the return of Arctic flowers to northern Europe at a time when the ice age had been over for more than a thousand years. Now those lake sediments seem to provide exemplary evidence of how the climate shifted and shifted again during that period. The reappearance of cold-loving beetles in the British Isles and the resurgence of tiny, cold-tolerant foraminifers in the North Atlantic can also be interpreted in these terms. And so too, arguably, can the rise of human civilization and, by extension, the progress of climatology.

One night, I was sitting in the geodesic dome at North GRIP with Steffensen. He was coming to the end of a month on the ice, and had the weatherbeaten look of someone who has spent too long at sea. "If you look at the paleoclimatic output of ice cores, it has really changed the picture of the world, our view of past climates, and of human evolution," he said, while, next to us, a group of graduate students played board games and listened to the soundtrack from *Buena Vista Social Club*. "Now you're able to put human evolution into a climatic framework. You can ask, Why did human beings not make civilization fifty thousand years ago? You know that they had just as big brains as we have today. When you put it in a climatic framework, you can say, Well, it was the ice age. And also this ice age was so climatically unstable that each time you had the beginning of a culture they had to move. Then comes the present interglacial — 10,000 years of very stable climate. The perfect conditions for agriculture. If you look at it, it's amazing. Civilizations in Persia, in China, and in India start at the same time, maybe 6,000 years ago. They all developed writing and they all developed religion and they all built cities, all at the same time, because the climate was stable. I think that if the climate would have been stable 50,000 years ago it would have started then. But they had no chance."

The only way into North GRIP is through Kangerlussuaq, and it is the only way out as well. The name means "very long fjord," and Kangerlussuaq does indeed lie at the end of a 180-mile-long fjord, which opens out into the Davis Strait. The setting is spectacular — snow-covered mountains rising out of a glacial plain. The town itself, however, is mostly poured concrete and corrugated iron, the remains of a now defunct American Air Force base that was called Sonderstrom, or, for short, Sondy. The night I arrived, I was invited to dinner at the town's best restaurant, at the airport. I missed the hors d'oeuvres, which had included whale skin, but arrived in time for the entrée, which was reindeer. When I left, at about 9 P.M., I saw a musk ox on the hillside just beyond the terminal.

The edge of the ice sheet lies some 10 miles away, and it can be seen — a ghostly white blur in the distance — by climbing just about any hill. After returning from North GRIP, I had a few days to spend in Kangerlussuaq, and one afternoon I hitched a ride out to the ice with some glaciologists who were also awaiting flights home. We took a dirt road that had been built by Volkswagen, and someone put Pink Floyd on the truck's CD player. Almost as soon as we got out of town, we were in the wild, cutting through fields of tiny purple Arctic rhododendron.

The Volkswagen road goes all the way up onto the ice sheet and ends 100 miles later at a test track. (Rumor has it that there is also a three-star hotel and restaurant in a modular building that was trucked out to the site.) We stopped far short of that, at a fast-running river, brown with silt. The ice sheet rose up beyond it, like a wall, 200 feet high. It was a startling shade of blue. One of the glaciologists explained that the color was an effect of the ice's peculiar density. Up at North GRIP, a set of poles that are slowly drifting apart mark the glacier's flow; at the edge of the ice, the same process produces more dramatic results. As we were talking, a huge section of the wall tore free and crashed into the river, sprinkling us with ice chips.

Although it was a clear blue day, a chill wind was blowing off the glacier, and, after we had all finished taking pictures, we climbed back into the truck. Soon we passed a small herd of reindeer that had come down to drink at a half-frozen lake, and then, a little later, the remains of a recent reindeer hunt — a pile of hooves with the fur still on them. The only other signs of human life we encountered were some ancient Inuit graves, or cairns; traditionally, Greenlanders buried their dead under mounds of rocks, a concession to the fact that most of the year the ground is frozen solid.

Humans are a remarkably resourceful species. We have spread into every region of the globe that is remotely habitable, and some, like Greenland, that aren't even that. The fact that we have managed this feat in an era of exceptional climate stability does not diminish the accomplishment, but it does make it seem that much more tenuous. As we drove back to Kangerlussuaq, listening to Pink Floyd — "Hey you, out there in the cold / Getting lonely, getting old / Can you feel me?" — I found myself thinking again about the Greenland Norse. They had arrived on the island at a moment of uncharacteristically benign weather, but they wouldn't have had any way of knowing this. Then the weather turned, and they were gone.

ANDREW LAWLER

Treasure Under Saddam's Feet

FROM *Discover*

YOU ARE DRIFTING down the sluggish, muddy Tigris River on a reed raft, headed for a prominent spur of rock rising from a broad plain. Upon the rock stand the massive walls of brightly painted temples. Just behind them soars a brilliantly colored temple tower, or ziggurat, nearly 200 feet high, with a pair of smaller ziggurats in the background. Beyond sprawl the roofs of vast royal palaces housing magnificent reception halls and sealed underground tombs.

As the boat docks, sunbaked sailors and stevedores unload goods and tribute, everything from African ivory to Anatolian metals to Afghan lapis lazuli. Traders, donkeys, pilgrims, horses, artisans, priests, and diplomats pass through the dozen gates above. This is bustling Assur, a town of perhaps 30,000, one of the most dazzling sights in Mesopotamia and in the entire ancient world.

Assur was the birthplace and spiritual center of Assyria, the mother of all empires. At its zenith in the seventh century B.C., Assyria's rule stretched from the southern borders of Egypt to the Persian Gulf and north to the Turkish highlands. Although largely forgotten, Assyrians assembled the first truly multicultural empire, built the first great library, and designed some of the first planned cities. They were the first to divide the circle into 360 degrees and gave the world technologies ranging from aqueducts to paved roads. The Assyrians also laid the foundation for the more famous Persian, Greek, Roman, and Parthian empires.

Today Assur is nothing more than a desolate mound. Countless seasons of rain and desert wind have eaten away at the mudbrick ziggurat, and nineteenth-century Ottoman barracks cover the once holy promontory. Nonetheless, this is a troubling site. Al-

though there is great promise of archaeological treasure beneath
the rubble here, the area faces even greater obliteration. The Iraqi
government is planning to complete a massive dam downstream
on the Tigris. Within four years, the ancient metropolis — the oldest and most revered site among a chain of Assyrian cities — will
become a muddy stump of an island in a vast lake. And Assur's hin-
terland — the cities and towns and villages that are buried nearby
— will be sunk, their wealth of artifacts left to dissolve. All of which
has the German archaeologist Peter Miglus in a state of despair. He
and his team have waited years, through the Gulf War and its aftermath, to resume digging at Assur. Now he looks sadly across the
Tigris valley and says: "This is the core of Assyria, and we have far
more questions than answers about life here."

Were a dam to threaten a well-known ancient site like Pompeii,
the international outcry would be compelling. But Iraq's status as
an international pariah, not to mention Assur's obscurity, has so far
doomed efforts to seek the empire's roots. The desperation among
Assyrian scholars over the impending loss is made only more acute
by the recent spectacular discovery of tombs in the newer Assyrian
capital of Nimrud. That find — which includes the skeletons of the
consorts to the most powerful Assyrian kings as well as caches of
finely worked gold and precious stones — rivals even the 1920s discovery of King Tut's tomb and the royal graves of Ur. The Nimrud
tombs, along with new texts, translations, and computer simulations of Assyrian palaces, provide a look at what might soon be lost
in Assur.

Assyrians appeared relatively late on the Mesopotamian stage —
around 2000 B.C. — by which time the great city-states of Sumer
and Babylonia had already emerged. By the thirteenth century
B.C., they had firmly established themselves as a regional power.
With the help of a growing professional military equipped with
swift horses, chariots, and iron swords and lances, Assyria secured
and expanded its trade routes. Paved roads — a novelty — provided easy transport year-round for traders and soldiers alike.

By 800 B.C., the lands under Assyrian control came to embrace a
far larger territory than any previous empire's. Assyria's great cities
— Assur, Nimrud (then known as Calah), Khorsabad, Nineveh —
were unrivaled in size and magnificence. Aqueducts watered gar-
dens for palaces covering grounds the size of a football field. Massive walls — stretching 7 miles long at Nineveh — protected tens

of thousands. But in 614 B.C., a coalition of Babylonians from the south and Medes from the Iranian plateau to the east swept through, laying waste to Assur and damaging Nimrud. Two years later, the combined armies destroyed Nimrud and laid siege to Nineveh; after the battle, Nineveh was burned.

Still, some ancient treasure remained. In 1988 the Iraqi archaeologist Muzahem Hussein noticed that bricks on the floor of a palace room at Nimrud looked out of place. While putting them back into position, he discovered that they were sitting on top of a vault. When he looked for an entrance, he found a vertical shaft and a stairway that led into a tomb. After two weeks of hauling out dust, he caught a glimpse of gold jewelry. "I couldn't believe my eyes," he recalled. Muzahem, a lean and quiet man who grew up in nearby Mosul, didn't then realize he had made one of the most spectacular discoveries in archaeological history.

By the time the Gulf War began in 1991, Muzahem had uncovered three additional tombs, each with its own collection of skeletons, gold jewelry, and personal items — the richest find from the ancient world since the heady days of the 1920s, when Howard Carter and Lord Carnarvon opened Tutankhamen's tomb in Egypt while Leonard Woolley excavated the royal graves in the southern Mesopotamian city of Ur.

"In terms of sheer spectacle, there has been nothing like this in Mesopotamian archaeology" since Woolley's finds, says Joan Oates, a British researcher who worked at the site in the 1950s along with Agatha Christie, who was married to the excavation's director. The finds include a finely wrought gold crown topped by delicately winged female figures, chains of tiny gold pomegranates, dozens of earrings of gold and semiprecious stones, even gold rosettes that decorated the dresses of the deceased.

The war, however, interrupted further study, and for the past decade Iraq's political position has made excavation nearly impossible. Then, last year, the government gave permission for foreign scholars to excavate. But any archaeologist working here must contend with much more than the blistering heat and biting flies. Armed looters roam the desert, and local archaeologists — those who didn't die in the Iran-Iraq war during the 1980s and who didn't flee in the aftermath of the Gulf War — routinely carry rifles at dig sites. And while most Iraqis treat scholars with great respect, some Western practices, such as photography, are looked upon with suspicion. This is a land where, in the words of one foreign archaeologist, "anyone with a camera is either a spy or stupid."

The importance of the sites in Iraq became public only this spring when Muzahem and other Iraqi archaeologists presented the contents of four tombs at a London conference. The first tomb held a still-sealed sarcophagus, with the remains of a woman about fifty years old and a collection of exquisite jewelry of gold and semiprecious stones. The second, found less than 300 feet away, proved more sensational. Two queens — consorts to kings rather than rulers in their own right — were laid to rest here, one on top of the other in the same sarcophagus, wrapped in embroidered linen and covered with gold jewelry, including a crown, a mesh diadem, 79 earrings, 30 rings, 14 armlets, 4 anklets, 15 vessels, and many chains.

The second tomb included a curse, threatening the person who opened the grave of Queen Yaba — the wife of powerful Tiglathpileser III (744–727 B.C.) — with eternal thirst and restlessness. The curse specifically warns against disturbing the tomb or placing another corpse in it. Strangely, despite this curse, the second corpse was added after Yaba's death. Forensic specialists determined that both women were thirty to thirty-five years old; the cause of death is not clear. But the evidence indicates that Yaba was buried first. At some later date — twenty to fifty years after the first interment — the second corpse was placed on top of the first.

On the upper body was a gold bowl with the inscription "Atalia, queen of Sargon, king of Assyria," who ruled from 721 to 705 B.C. Another bowl mentions "Banitu, queen of Shalmaneser V," who ruled from 726 to 722 B.C. Because the second corpse was placed in the sarcophagus last, researchers assume the remains are those of Atalia. But what of Banitu? An alabaster jar in the tomb contains organic material that some archaeologists suspect may be Banitu's remains.

The Oxford scholar Stephanie Dalley proposes an explanation for the two corpses and three names. She suggests that Banitu and Yaba are the same woman — *yaba* being a Western Semitic word meaning "beautiful," while Banitu is a name in Akkadian, the language from which Assyrian is derived. Moreover, Atalia may be a Western Semitic name, indicating that both women may have been foreigners married to the Assyrian king. The theory remains controversial with scholars.

Atalia's presence poses an additional riddle: The body was apparently dried or smoked at temperatures of 300 to 500 degrees Fahrenheit for several hours. This could have been a burial practice or an effort to preserve a body for a long trip. Whatever its function, it provides the first evidence of mummification in ancient Mesopotamia.

A third tomb, uncovered in 1989, is even more mysterious. The main room had been robbed in antiquity, but an inscription named it as the resting place of Mullissu-mukannisat Ninua, queen of Ashurnasirpal II and mother of Shalmaneser III. The grave robbers missed the antechamber, packed with three bronze coffins containing human remains and jewelry. One contained the bones of six people, including a young adult, three children, a baby, and a fetus. A second coffin contained a young woman — most likely a queen, given the magnificent gold crown she wore — as well as a child. A third coffin held five adults, including a man fifty-five to sixty-five years old in unusually good physical condition at the time of his death. A golden vessel with the name of Samsuilu, an illustrious field marshal who served under at least three kings, was found in the third coffin. Some, if not all, of the bones in the coffin appear to have been buried elsewhere and then reinterred together later. Why and when remain a mystery. Multiple burials are not common in Assyria.

Fearing looters would get wind of the finds, Iraqi archaeologists had to excavate so quickly that fragile clues such as textiles and pollen were lost. But German forensic specialists, working with what is left of the human remains, have turned up some hints about the health of royal Assyrians.

The five adults with dental remains had healthy teeth, probably reflecting the better nutrition and softer foods available at the top of the Assyrian social structure. Only one, Atalia, suffered from cavities. Yaba and Atalia, however, also suffered from dental abscesses at some point in their short lives. In addition, all the adults suffered from chronic sinus infections.

Five out of eight skeletons showed signs of health problems ranging from high fevers and infections to poor nutrition. And out of the seven skeletons that could be studied for changes in the skull, six — including those of Yaba and Atalia — showed telltale areas of thickened skull, indicating they had survived a bout with meningitis. "The Assyrian queens have just begun to speak to us," says Michael Müller-Karpe, a German archaeologist. "And we are looking forward to more answers — especially to those which can be expected from DNA analyses." That will include finding out if they were the daughters of distant kings or native royal Assyrians. Or if Atalia is the daughter of Yaba.

For millennia, the Assyrians have been remembered through the legends of their enemies. The biblical prophet Isaiah railed against "the king of Assyria's boastful heart, and his arrogant insolence." The prophet Nahum speaks of the "unrelenting cruelty" of Assyrian leaders. And the second Book of Kings warns that "the kings of Assyria have exterminated all the nations, they have thrown their gods on the fire." According to John Malcolm Russell, an art historian and archaeologist at the Massachusetts College of Art, "it's like a history of the United States written by the Ayatollah Khomeini."

Yet what British and French explorers found nearly two millennia later seemed to confirm that image. Stone friezes from Nimrud and Nineveh depict war chariots trundling over the bodies of enemy soldiers, women and children deported from their homes, and an Assyrian king and his queen relaxing over wine and fruit in a verdant garden while an enemy leader's head swings from a tree nearby. The repetitive carvings of muscled, bearded, and warmongering princes that appear on the friezes have remained the best-known emblem of Assyrian society.

Russell, however, views the images as carefully positioned propaganda. While working at Nimrud and Nineveh in the 1980s, he noted that images of plunder, brutality, and war are reserved largely for the reception and throne rooms, where foreign diplomats and leaders met the Assyrian king. "The reliefs are at their shrillest in the public rooms," he says.

In rooms reserved for the king and his retinue, the walls are covered with less intimidating figures. These emblems, says Russell, may be designed to ward off evil spirits. In the king's own bedchambers there are no images at all, merely cuneiform inscriptions asserting the ruler's sovereignty. Russell speculates that the writing may have served as a protective talisman for a vulnerable Assyrian leader. A few rulers, like Sennacherib, were known to have died at the hands of relatives in palace coups.

"I don't think the Assyrians were any more bloodthirsty than their contemporaries," says Nicholas Postgate, a professor of Assyriology at Cambridge University. "Mind you, I would rather not have been on the other side."

Like those who ruled the Roman Empire, Assyrian kings welcomed subjects who were willing to become part of the empire. Those who resisted were conquered. The men were often killed, while the women and children were sometimes abducted and relocated to distant regions. Joan Oates and her archaeologist husband, David, have found tablet inscriptions that say displaced civilians were equipped with food, oil, clothes, and shoes. Refugees were also encouraged to marry. "Under the Assyrians, the entire area became a vast experiment in cultural mixing," writes the Washington State University historian Richard Hooker.

The excitement among Assyrian scholars about the reopening of Iraq to archaeological excavation is tempered by concern about the damage to Assur should the dam be completed. Iraqi officials have discussed building a giant wall to surround the site or taking steps to prevent the waters from rising above a certain height, but Peter Miglus is skeptical.

"You can't save Assur if it's in the vicinity of a dam," he says. The clay underneath Assur will wick up water, he believes, destroying what lies below, even if the surface is above water.

That solution also ignores dozens of other sites in the valley never examined by archaeologists. The best that can be hoped for, says Miglus, is a quick Iraqi call for international help or that senior Iraqi officials — perhaps Saddam Hussein himself — will halt or delay the effort. The minister of higher education and scientific research, Humam Abdul Khaliq A. Ghafour, backs the creation of an Assyrian research center in Mosul to draw international scholars and encourage a new generation of Iraqi researchers. Drowning Assur could prove internationally embarrassing. "We will do our best to hinder, or at least delay, the inauguration of this [dam] project," he said recently in his Baghdad office. "We don't want the slightest damage to Assur."

The dam, however, is under the control of the powerful Irrigation Ministry, and work is well under way. Foreign help is unlikely, given the growing fears that the United States will wage war against Saddam. All Miglus can do is wait and organize another season of digging before the waters rise.

DANIEL LAZARE

False Testament

FROM *Harper's Magazine*

NOT LONG AGO, archaeologists could agree that the Old Testament, for all its embellishments and contradictions, contained a kernel of truth. Obviously, Moses had not parted the Red Sea or turned his staff into a snake, but it seemed clear that the Israelites had started out as a nomadic band somewhere in the vicinity of ancient Mesopotamia; that they had migrated first to Palestine and then to Egypt; and that, following some sort of conflict with the authorities, they had fled into the desert under the leadership of a mysterious figure who was either a lapsed Jew or, as Freud maintained, a high-born priest of the royal sun god Aton, whose cult had been overthrown in a palace coup. Although much was unknown, archaeologists were confident that they had succeeded in nailing down at least these few basic facts.

That is no longer the case. In the last quarter century or so, archaeologists have seen one settled assumption after another concerning who the ancient Israelites were and where they came from proved false. Rather than a band of invaders who fought their way into the Holy Land, the Israelites are now thought to have been an indigenous culture that developed west of the Jordan River around 1200 B.C. Abraham, Isaac, and the other patriarchs appear to have been spliced together out of various pieces of local lore. The Davidic Empire, which archaeologists once thought as incontrovertible as the Roman, is now seen as an invention of Jerusalem-based priests in the seventh and eighth centuries B.C. who were eager to burnish their national history. The religion we call Judaism does not reach well back into the second millennium B.C. but appears to be, at most, a product of the mid-first.

This is not to say that individual elements of the story are not older. But Jewish monotheism, the sole and exclusive worship of an ancient Semitic god known as Yahweh, did not fully coalesce until the period between the Assyrian conquest of the northern Jewish kingdom of Israel in 722 B.C. and the Babylonian conquest of the southern kingdom of Judah in 586.

Some twelve to fourteen centuries of "Abrahamic" religious development, the cultural wellspring that has given us not only Judaism but Islam and Christianity, have thus been erased. Judaism appears to have been the product not of some dark and nebulous period of early history but of a more modern age of big-power politics in which every nation aspired to the imperial greatness of a Babylon or an Egypt. Judah, the sole remaining Jewish outpost by the late eighth century B.C., was a small, out-of-the-way kingdom with little in the way of military or financial clout. Yet at some point its priests and rulers seem to have been seized with the idea that their national deity, now deemed to be nothing less than the king of the universe, was about to transform them into a great power. They set about creating an imperial past commensurate with such an empire, one that had the southern heroes of David and Solomon conquering the northern kingdom and making rival kings tremble throughout the known world. From a "henotheistic" cult in which Yahweh was worshiped as the chief god among many, they refashioned the national religion so that henceforth Yahweh would be worshiped to the exclusion of all other deities. One law, that of Yahweh, would now reign supreme.

This is not, of course, the story that we have all been led to believe is, at least to some degree, history. This is not the story told, for instance, in such tomes as Paul Johnson's 1987 bestseller, *A History of the Jews,* from which we learn that Abraham departed the ancient city of Ur early in the second millennium B.C. as part of a great westward trek of "Habiru" (i.e., Hebrew) nomads to the land of Canaan. "[T]hough the monotheistic concept was not fully developed in [Abraham's] mind," Johnson writes, "he was a man striving towards it, who left Mesopotamian society precisely because it had reached a spiritual impasse." Now, however, we know that this statement is mainly bosh. Not only is there no evidence that any such figure as Abraham ever lived, but archaeologists believe that there is no way such a figure could have lived given what we now know about ancient Israelite origins.

A few pages later, Johnson declares that "we can be reasonably sure that the Exodus occurred in the thirteenth century B.C. and had been completed by about 1225 B.C." Bosh as well. A growing volume of evidence concerning Egyptian border defenses, desert sites where the fleeing Israelites supposedly camped, etc., indicates that the flight from Egypt did not occur in the thirteenth century before Christ; it never occurred at all. Although Johnson writes that the story of Moses had to be true because it "was beyond the power of the human mind to invent," we now know that Moses was no more historically real than Abraham before him. Although Johnson adds that Joshua, Moses's lieutenant, "began and to a great extent completed the conquest of Canaan," the Old Testament account of that conquest turns out to be fictional as well. And although Johnson goes on to inform his readers that after bottling up the Philistines in a narrow coastal strip, King David "then moved east, south and north, establishing his authority over Ammon, Moab, Edom, Aram-Zobar and even Aram-Damascus in the far north-east," archaeologists believe that David was not a mighty potentate whose power was felt from the Nile to the Euphrates but rather a freebooter who carved out what was at most a small duchy in the southern highlands around Jerusalem and Hebron. Indeed, the chief disagreement among scholars nowadays is between those who hold that David was a petty hilltop chieftain whose writ extended no more than a few miles in any direction and a small but vociferous band of "biblical minimalists" who maintain that he never existed at all.

In classic Copernican fashion, a new generation of archaeologists has taken everything its teachers said about ancient Israel and stood it on its head. Two myths are being dismantled as a consequence: one concerning the origins of ancient Israel and the other concerning the relationship between the Bible and science. Back in the days when archaeology was buttressing the old biblical tales, the relationship between science and religion had warmed considerably; now the old chill has crept back in. The comfy ecumenicism that allowed one to believe in, say, modern physics and Abraham,

Isaac, et al., is disappearing, replaced by a somewhat sharper dividing line between science and faith. The implications are sweeping
— after all, it is not the Song of the Nibelungen or the Epic of
Gilgamesh that is being called into question here but a series of
foundational myths to which fully half the world's population, in
one way or another, subscribes.

So how did such a glorious revolution come to be? As is usually
the case, we must first look to when cracks started developing in
the *ancien régime.*

Ironically, the new archaeology represents something of a circling
back to what was once known as the "Higher Criticism," a largely
German school of biblical study that relied solely on linguistic and
textual analysis. By the late nineteenth century, members of this
school had arrived at the conclusion that the first five books of the
Old Testament — variously known as the Five Books of Moses, the
Torah, or the Pentateuch — were not written by Moses himself, as
tradition would have it. Rather, they were largely products of a
"post-exilic period" in which Jewish scribes, newly released from
captivity in Babylon, set about putting a jumbled collection of ancient writings into some sort of coherent order. The Higher Criticism did not topple the Old Testament as a whole, but it did conclude that Abraham, Isaac, and the other tribal founders depicted
in the Book of Genesis were no more real than the heroes of Greek
or Norse mythology. As the German scholar Julius Wellhausen put
it in the 1870s: "The whole literary character and loose connection
of the . . . story of the patriarchs reveal how gradually its different
elements were brought together, and how little they have coalesced
into a unity." Rather than a chronicle of genuine events, the history
that Genesis set forth was an artificial construct, a narrative framework created long after the facts in order to link together a series of
unconnected folktales like pearls on a string.

If the linguists of the Higher Criticism were generally skeptical in
regard to the Old Testament, modern biblical archaeology as it began taking shape in the early nineteenth century was something
entirely different. The first modern archaeologists to set foot in the
Holy Land were New England Congregationalists, determined to
make use of rigorous scientific methods in order to strip away centuries of what they regarded as Roman Catholic superstition and

prejudice. As the American biblical scholar Edward Robinson, who first came to Palestine in 1838, put it, he would accept nothing until it was absolutely proven. And yet, as a dutiful Calvinist, Robinson assumed from the outset that whatever he uncovered would broadly confirm what he had learned years earlier in Sunday school. Evidence that buttressed the biblical account was eagerly sought out while evidence that contradicted it was ignored. British archaeologists set sail a generation later with an even more explicit set of preconceptions. As the archbishop of York told the newly created Palestine Exploration Fund in London in 1865,

> This country of Palestine belongs to *you* and to *me*, it is essentially ours. It was given to the Father of Israel [i.e., Abraham] in the words: "Walk through the land in the length of it, and in the breadth of it, for I will give it unto thee." *We* mean to walk through Palestine in the length and in the breadth of it, because that land has been given unto us . . .

The first archaeologists were thus guilty of one of the most elementary of scientific blunders: rather than allowing the facts to speak for themselves, they tried to fit them into a preconceived theoretical framework. Another layer of political mystification was added in the twentieth century by Zionist pioneers, eager for evidence that the Jewish claim to the Holy Land was every bit as ancient as the Old Testament said it was. In 1928 members of a settlement known as Beth Alpha uncovered an ancient synagogue mosaic while digging an irrigation ditch. Since the settlers were members of a left-wing faction known as Hashomer Hatzair, it was inevitable that some would argue that the find should be left to the dustbin of history and that the work of building a modern agricultural settlement should continue uninterrupted. But others recognized its significance: the more evidence they uncovered of an ancient Jewish presence in the Holy Land, the more they would succeed in legitimizing a modern colonization effort. As the number of digs multiplied and turned into a national passion, what the Israeli archaeologist Eliezer Sukenik described as a specifically "Jewish archaeology" was born.

The result was a happy union of science, religion, and politics that by the 1950s would eventually bring together everyone from Christian fundamentalists in the American heartland to the Israeli military establishment. When David Ben-Gurion, the founder of modern Israel, spoke of a sweeping offensive in the 1948 War of

Independence, he did so in language purposely evocative of the Book of Joshua. The armies of Israel, he declared, had "struck the kings of Lod and Ramleh, the kings of Beit Naballa and Deir Tarif, the kings of Kola and Migdal Zedek . . ." Yigael Yadin, Eliezer Sukenik's son, who was not only Israel's leading archaeologist but a top military commander, referred to an Israeli military incursion into the Sinai by quipping that it was the first time Israeli forces had set foot on the peninsula in 3,400 years. All assumed that the ancient events Israel claimed to be reenacting had actually occurred.

The politicization of archaeology reached something of a climax in the early 1960s, when Yadin was put in command of the excavation of Masada, a hilltop fortress where nearly 1,000 Jewish warriors had committed suicide rather than surrender to the Romans in A.D. 73. In Yadin's hands, Masada emerged as Israel's preeminent nationalist shrine, a place where military recruits were assembled to take an oath of allegiance in dramatic nighttime ceremonies — this despite complaints on the part of a few scholars that evidence for a mass suicide was lacking and that there was reason to believe that ancient accounts of the event were deliberately falsified.

Around this time, the pop novelist James Michener summed up the state of official belief in his heavy-breathing bestseller *The Source* (which this writer savored as a teenager). Using a fictional archaeological dig to weave a series of tales about Palestinian life from prehistoric times to the modern era, Michener briskly laid out the middlebrow orthodoxy of the day: i.e., that God had entered into a pact with the ancient Israelites early in the second millennium B.C., that Jews had dominated the Holy Land for some 2,000 years thereafter, and that with the birth of modern Israel they were claiming their birthright. "Deuteronomy is so real to me," Michener has a fictional Israeli archaeologist declare, "that I feel as if my immediate ancestor — say, my great-grandfather with desert dust still in his clothes — came down that valley with goats and donkeys and stumbled onto this spot." Michener says of another fictional archaeologist, an American who has just been reading the Torah,

This time he gained a sense of the enormous historicity of the book . . . He now read the Ten Commandments as if he were among the tribes listening to Moses. It was he who was coming out of Egypt, dying of thirst in the Sinai, retreating in petulant fear from the first invasion of the Promised Land. He put the Bible down with a distinct sense of having read the history of a real people . . .

Yet it was precisely this "historicity" that was beginning to come under fire. Resurrecting a theory first proposed in the 1920s, an Israeli named Yohanan Aharoni infuriated the Israeli archaeological establishment by arguing that evidence in support of an Israelite war of conquest in the thirteenth century B.C. was weak and unconvincing. Basing his argument on a redating of pottery shards found at a dig in the biblical city of Hazor, Aharoni proposed instead that the first Hebrew settlers had filtered into Palestine in a nonviolent fashion, peacefully settling among the Canaanites rather than putting them to the sword. Although archaeologists claimed in the 1930s to have uncovered evidence that the walls of Jericho had fallen much as the Book of Joshua said they had, a British archaeologist named Kathleen Kenyon was subsequently able to demonstrate, based on Mycenaean pottery shards found amid the ruins, that the destruction had occurred no later than 1300 B.C., seventy years or more before the conquest could have happened. Whatever caused the walls of Jericho to come tumbling down, it was not Joshua's army.

The enormous ideological edifice that Yigael Yadin and others had erected was weakening at the base. Whereas formerly every pottery fragment or stone tablet appeared to confirm the biblical account, now nothing seemed to fit. Attempting to pinpoint precisely when Abraham had departed the ancient city of Ur, the American scholar William F. Albright, a pillar of the archaeological establishment until his death in 1971, theorized that he had left as part of a great migration of "Amorite" (literally "western") desert nomads sometime between 2100 and 1800 B.C. This was the theory that Paul Johnson would later cite in *A History of the Jews*. Subsequent research into urban development and nomadic growth patterns indicated that no such mass migration had taken place and that several cities mentioned in the Genesis account did not exist during the time frame Albright had suggested. Efforts to salvage the theory by moving up Abraham's departure to around 1500 B.C. foundered when it was pointed out that, this time around, Genesis failed to mention cities that *did* dominate the landscape during this period. No matter what time frame was advanced, the biblical text did not accord with what archaeologists were learning about the land of Canaan in the second millennium.

This was not all. As Israel Finkelstein, an archaeologist at Tel Aviv University, and Neil Asher Silberman, a journalist who specializes in biblical and religious subjects, point out in their recent book, *The Bible Unearthed,* the patriarchal tales make frequent mention of camel caravans. When, for example, Abraham sent one of his servants to look for a wife for Abraham's son, Isaac, Genesis 24 says that the emissary "took ten of his master's camels and left, taking with him all kinds of good things from his master." Yet analysis of ancient animal bones confirms that camels were not widely used for transport in the region until well after 1000 B.C. Genesis 26 tells of Isaac seeking help from a certain "Abimelech, king of the Philistines." Yet archaeological research has confirmed that the Philistines were not a presence in the area until after 1200 B.C. The wealth of detail concerning people, goods, and cities that makes the patriarchal tales so vivid and lifelike, archaeologists discovered, was reflective of a period long after the one that Albright had pinpointed. These details were reflective of the mid-first millennium, not the early second.

In hindsight, it all seems so obvious. An ancient text purporting to be a record of events centuries earlier — how could it not fall short of modern historical standards? How could it not reflect contemporary events more than events in the distant past? Beginning in the 1950s, doubts concerning the Book of Exodus multiplied just as they had about Genesis. The most obvious concerned the complete silence in contemporary Egyptian records concerning the mass escape of what the Bible says were no fewer than 603,550 Hebrew slaves. Such numbers no doubt were exaggerated. Yet considering how closely Egypt's eastern borders were patrolled at that time, how could the chroniclers of the day have failed to mention what was still likely a major security breach?

Old-guard academics professed to be untroubled. John Bright, a prominent historian, was dismissive of the entire issue. "Not only were Pharaohs not accustomed to celebrate reverses," he wrote in *A History of Israel,* long considered the standard account, "but an affair involving only a party of runaway slaves would have been to them of altogether minor significance." The scribes' silence concerning the mysterious figure of Moses, Bright went on, was also of no account. Regardless of what the chronicles did or did not say, "The events of exodus and Sinai require a great personality behind them. And a faith so unique as Israel's demands a founder as surely as does Christianity — or Islam, for that matter."

This was dogma masquerading as scholarship. Not only was there a dearth of physical evidence concerning the escape itself, as archaeologists pointed out, but the slate was blank concerning the nearly five centuries that the Israelites had supposedly lived in Egypt prior to the Exodus as well as the forty years that they supposedly spent wandering in the Sinai. Not so much as a skeleton, campsite, or cooking pot had turned up, Finkelstein and Silberman noted, even though "modern archeological techniques are quite capable of tracing even the very meager remains of hunter-gatherers and pastoral nomads all over the world." Indeed, although archaeologists have found remains in the Sinai from the third millennium B.C. and the late first, they have found none from the thirteenth century.

As with Abraham, the effort to nail down a time frame for the departure created more problems than it resolved. Archaeologists had long zeroed in on a relatively narrow window of opportunity in the thirteenth century B.C. bounded by two independently verifiable events — the start of work on two royal cities in which the Book of Exodus says Hebrew slaves were employed ("and they built Pithom and Rameses as store cities for Pharaoh") and the subsequent erection of a victory stele, or monument, that describes a people identified as "Israel" already existing in Canaan. Hence, the flight into the Sinai had to have taken place either during the reign of a pharaoh known as Rameses or shortly after the death of Ramses II in 1213 B.C.

Once again the theory didn't add up. The Book of Numbers states that, following their escape, the Israelites came under attack from the "Canaanite king of Arad, who lived in the Negev," as they were "coming along the road to Atharim." But although excavations showed that a city of Arad existed in the early Bronze Age from roughly 3500 to 2200 B.C., and that an Iron Age fort arose on the site beginning in roughly 1150 B.C., it was deserted during the years in between. The Pentateuch says the Hebrews did battle with

Sihon, king of the Amorites, at a city called Heshbon, but excavations have revealed that Heshbon did not exist during this period either. Nor did Edom, against whose king the Old Testament says the ancient Jews also made war.

Then came a series of archaeological studies conducted in the aftermath of the Six-Day War in 1967. Previously, archaeologists had intensively studied specific sites and locales, digging deep in order to determine how technology and culture had changed from one century to the next. Now they tramped through hills and valleys looking for pottery shards and remnants of ancient walls in order to map out how settlement patterns had ebbed and flowed across broad stretches of terrain. Whereas previously archaeologists had concentrated on the lowland cities where the great battles mentioned in the Bible were said to have taken place, they now shifted their attention to the highlands located in the present West Bank. The results were little short of revolutionary. Rather than revealing that Canaan was entered from the outside, analysis of ancient settlement patterns indicated that a distinctive Israelite culture arose locally around 1200 B.C. as nomadic shepherds and goatherds ceased their wanderings and began settling down in the nearby uplands. Instead of an alien culture, the Israelites were indigenous. Indeed, they were highly similar to other cultures that were emerging in the region around the same time — except for one thing: whereas archaeologists found pig bones in other sites, they found none among the Israelites. A prohibition on eating pork may have been one of the earliest ways in which the Israelites distinguished themselves from their neighbors.

Thus there was no migration from Mesopotamia, no sojourn in Egypt, and no exodus. There was no conquest upon the Israelites' return and, for that matter, no peaceful infiltration such as the one advanced by Yohanan Aharoni. Rather than conquerors, the Hebrews were a native people who had never left in the first place. So why invent for themselves an identity as exiles and invaders? One reason may have been that people in the ancient world did not establish rights to a particular piece of territory by farming or by raising families on it but by seizing it through force of arms. Indigenous rights are an ideological invention of the twentieth century A.D. and are still not fully established in the twenty-first, as the plight of today's Palestinians would indicate. The only way that the

Israelites could establish a moral right to the land they inhabited was by claiming to have conquered it sometime in the distant past. Given the brutal power politics of the day, a nation either enslaved others or was enslaved itself, and the Israelites were determined not to fall into the latter category.

If the Old Testament is to be believed, David and Solomon, rulers of the southern kingdom of Judah from about 1005 to about 931 B.C., made themselves masters of the northern kingdom of Israel as well. They represent, in the official account, a rare moment of national unity and power; under their reign, the combined kingdom was a force throughout the Fertile Crescent. The unified kingdom is said to have split into two rump states shortly after Solomon's death and, thus weakened, was all too easy for the Assyrian Empire and its Babylonian successor to pick off. But did a united monarchy encompassing all twelve tribes ever truly exist?

According to the Bible, Solomon was both a master builder and an insatiable accumulator. He drank out of golden goblets, outfitted his soldiers with golden shields, maintained a fleet of sailing ships to seek out exotic treasures, kept a harem of 1,000 wives and concubines, and spent thirteen years building a palace and a richly decorated temple to house the Ark of the Covenant. Yet not one goblet, not one brick, has ever been found to indicate that such a reign existed. If David and Solomon had been important regional power brokers, one might reasonably expect their names to crop up on monuments and in the diplomatic correspondence of the day. Yet once again the record is silent. True, an inscription referring to "Ahaziahu, son of Jehoram, king of the House of David" was found in 1993 on a fragment dating from the late ninth century B.C. But that was more than a hundred years after David's death, and at most all it indicates is that David (or someone with a similar name) was credited with establishing the Judahite royal line. It hardly proves that he ruled over a powerful empire.

Moreover, by the 1970s and 1980s a good deal of countervailing evidence — or, rather, lack of evidence — was beginning to accumulate. Supposedly, David had used his power base in Judah as a springboard from which to conquer the north. But archaeological surveys of the southern hill country show that Judah in the eleventh and tenth centuries B.C. was too poor and backward and sparsely populated to support such a military expedition. Moreover, there was no evidence of wealth or booty flowing back to the southern power base once the conquest of the north had taken place. Jerusalem seems to have been hardly more than a rural village when Solomon was reportedly transforming it into a glittering capital. And although archaeologists had long credited Solomon with the construction of major palaces in the northern cities of Gezer, Hazor, and Megiddo (better known as the site of Armageddon), recent analysis of pottery shards found on the sites, plus refined carbon-14 dating techniques, indicate that the palaces postdate Solomon's reign by a century or more.

Finkelstein and Silberman concluded that Judah and Israel had never existed under the same roof. The Israelite culture that had taken shape in the central hill country around 1200 B.C. had evolved into two distinct kingdoms from the start. Whereas Judah remained weak and isolated, Israel did in fact develop into an important regional power beginning around 900 B.C. It was as strong and rich as David and Solomon's kingdom had supposedly been a century earlier, yet it was not the sort of state of which the Jewish priesthood approved. The reason had to do with the nature of the northern kingdom's expansion. As Israel grew, various foreign cultures came under its sway, cultures that sacrificed to gods other than Yahweh. Pluralism became the order of the day: the northern kings could manage such a diverse empire only by allowing these cultures to worship their own gods in return for their continued loyalty. The result was a policy of religious syncretism, a theological pastiche in which the cult of Yahweh coexisted alongside those of other Semitic deities.

When the northern kingdom fell to the Assyrians, the Jewish priesthood concluded not that Israel had played its cards badly in the game of international politics but that by tolerating other cults it had given grave offense to the only god that mattered. Joining a stream of refugees to the south, the priests swelled the ranks of an influential political party dedicated to the proposition that the only way for Judah to avoid a similar fate was to cleanse itself of all rival beliefs and devote itself exclusively to Yahweh.

"They did wicked things that provoked Yahweh to anger. They worshiped idols, though Yahweh had said, 'You shall not do this.'" Such was the "Yahweh-alone" movement's explanation for Israel's downfall. The monotheistic movement reached a climax in the late seventh century B.C. when a certain King Josiah took the throne and gave the go-ahead for a long-awaited purge. Storming through the countryside, Josiah and his Yahwist supporters destroyed rival shrines, slaughtered alien priests, defiled their altars, and ensured that henceforth even Jewish sacrifice would take place exclusively in Jerusalem, where the priests could exercise tight control. The result, the priests and scribes believed, was a national renaissance that would soon lead to the liberation of the north and a similar cleansing there as well.

But then: disaster. After allowing his priests to establish a rigid religious dictatorship, Josiah rode off to rendezvous with an Egyptian pharaoh named Necho in the year 609 B.C. Although Chronicles says that the two monarchs met to do battle, archaeologists, pointing out that Josiah was in no position to challenge the mighty Egyptian army, suspect that Necho merely summoned Josiah to some sort of royal parley and then had him killed for unknown reasons. A model of pious rectitude, Josiah had done everything he thought God wanted of him. He had purified his kingdom and consecrated his people exclusively to Yahweh. Yet he suffered regardless. Judah entered into a period of decline, culminating some twenty-three years later in the Babylonian conquest and exile.

Does this mean that monotheism was nothing more than a con, a ruse cooked up by ambitious priests in order to fool a gullible population? As with any religion, cynicism and belief, realpolitik and genuine fervor, all came together in a way that we can barely begin to untangle. To say that the Jerusalem priesthood intentionally cooked up a phony history is to assume that the priests possessed a modern concept of historical truth and falsehood, and surely this is not so. As the biblical minimalist Thomas L. Thompson has noted, the Old Testament's authors did not subscribe to a sequential chronology but to some more complicated arrangement in which the great events of the past were seen as taking place in some foggy time before time. The priests, after all, were not inventing a past; they were inventing a present and, they trusted, a future.

Monotheism was unquestionably a great leap forward. At a time when there was no science, no philosophy, and no appreciable knowledge of the outside world, an obscure, out-of-the-way people

ATTENTION:

DATE SENT:

12-16

FROM:

ROUTE:

SOUTHEAST

NOTE:

DELIVERY PROVIDED BY:
MASSACHUSETTS REGIONAL
LIBRARY SYSTEMS

somehow conceived of a lone deity holding the entire universe in his grasp. This was no small feat of imagination, and its consequences were enormous. Monotheism's attempt at a unified field theory — a single explanation for everything from the creation of the universe to the origin of law — failed, but in failing it ensured that people would try doubly hard to come up with some new "theory of everything" to take its place. The monotheistic revolution continued to build because it enlisted a larger and larger portion of the population in its great totalizing effort. The Book of Kings tells of the discovery, during Josiah's reign, of a sacred book, filled with rules and regulations that the Jews had so far failed to follow, deep within the recesses of the Temple. In other cultures, the king might have huddled over the book with his advisers and priests. But not Josiah. He

> called together all the elders of Judah and Jerusalem. He went up to the temple of Yahweh with the men of Judah, the people of Jerusalem, the priests and the prophets — all the people from the least to the greatest. He read in their hearing all the words of the Book of the Covenant, which had been found in the temple of Yahweh. The king stood by the pillar and renewed the covenant in the presence of Yahweh . . . Then all the people pledged themselves to this covenant.

This was all quite novel. Whereas formerly the king and the priests alone were responsible to the national deity, now "all the people from the least to the greatest" took the pledge. The people had been transformed from mere onlookers into active participants. Arguably, the people of Judah were less free as a consequence of Josiah's reforms. Under the old pluralistic order they could sacrifice to other gods, and now they could sacrifice to just one. Yet with the new system's responsibilities to uphold the sacred covenant came the makings of a voice. No longer could the masses be counted on to remain silent.

Was the purpose of all this merely to pluck one tiny nation out of obscurity and elevate it above all others? If the Yahwists were groping for some concept of ethics to go with their universalism, for the most part they seem to have fallen woefully short. To quote Julius Wellhausen on the Jewish scriptures: "Monotheism is worked out to its furthest consequences, and at the same time is enlisted in the service of the narrowest selfishness."

A single, all-powerful god required a single set of sacred texts, and the process of composition and codification that led to what we now know as the Bible began under King Josiah and continued well into the Christian era. "Canonization" of this sort concentrated rather than dispelled questions of nationalism and universalism. A framework for faith, the Bible was equally a machine for generating heresy and doubt, and out of this debate eventually arose Christianity, Islam, Protestantism, and a great deal else besides.

The new universalism had enormous energy, encompassing as it did the entire cosmos and enlisting the entire population, but the new democratic spirit ran aground over the issue of universalism versus narrow nationalism. What, after all, was the point of mobilizing such a broad population in this manner? So that they could slaughter their neighbors all the more thoroughly? How could Moses prohibit murder and then, in Numbers 31, fly into a rage because a returning Israelite war party has slaughtered only the adult male Midianites? ("Now kill all the boys," he tells them when he calms down. "And kill every woman who has slept with a man, but save for yourselves every girl who has never slept with a man.") Was murder a crime only when it involved members of the in-group? Or was it a crime when it involved human beings in general, regardless of nationality? Did an emerging concept of a more equitable social order apply only to Israel or to other nations as well?

In one form or another, these questions have been with us ever since.

ELIZABETH F. LOFTUS

Memory Faults and Fixes

FROM *Issues in Science and Technology*

THE SEX ABUSE SCANDAL enveloping the Catholic Church has prompted vigorous calls for action: The Church should hand over to prosecutors a list of all its priests who have ever been accused in the past of sexual abuse; priests should be forced to resign if there has ever been an accusation; courts should devise ways to interpret laws that would allow criminal charges against priests even when the statute of limitation stands in the way; and Catholic bishops should be sued for violating federal antiracketeering laws — the laws that were intended to help dismantle Mafia-run organizations.

No one can fail to be moved by the anguished looks and words of those who recount tales of abuse by priests. But before we rush to adopt the called-for measures, we should look closely at recent news about overturned convictions in the courts and at the growing body of research about human memory. For centuries we have had experience with people who come to court to testify and take the familiar solemn oath. In light of what I have learned about human memory, I propose a more realistic alternative: "Do you swear to tell the truth, the whole truth, or whatever it is you think you remember?"

One has only to look at the growing number of cases in which DNA evidence has been used to exonerate innocent people. This year saw the release of the hundredth person nationwide to be freed from prison after genetic testing. Larry Mayes of Indiana, now fifty-two years old, spent twenty-one years in prison for a rape of a gas station cashier. The victim had failed to identify him in two separate lineups and picked him out only after she was hypnotized by police. Mayes's story is a common one; analyses of these DNA exoneration cases reveal that faulty eyewitness memory is the major cause of wrongful convictions.

Issues have also cropped up in cases that are built on the soggy foundation of "repressed" memory. The Arizona pediatrician John Danforth faced accusations by a former patient, Kim Logerquist, who suddenly remembered after an interval of two decades that he had repeatedly sexually molested her when she was between eight and ten years old. Her memories included a time when, after an assault, her panties were soaked with blood and she tossed them in the garbage can. At one point Logerquist wanted $3 million to $5 million in damages. Logerquist had been hospitalized fifty-seven times in the three years before her "flashbacks," memories that she claimed were repressed until triggered by viewing a television ad for children's aspirin. It is worth noting that Logerquist spent scores of hours in therapy in which she was urged to try to remember abuse that might explain her problems such as self-mutilation, depression, suicide attempts, obesity, and bulimia. Although she periodically denied it, records showed that she often spent time considering which men other than Danforth had abused her. A forensic psychiatrist bolstered Logerquist's story with the unsubstantiated claim that people who have flashbacks do not later produce inaccurate recollections of those events. Nothing could be further from the truth. Danforth, in his late sixties, steadfastly maintained his innocence and was eventually cleared. It took the last jury less than forty minutes to find for Danforth, to the delight of his extended family. The loud cheers were not surprising, coming from a family that had endured ten years of litigation as this landmark repressed memory case worked its way through various trials and appeals.

Thousands of cases based on recovered memory captured public attention throughout the 1990s. Some involved highly implausible or impossible memory claims such as intergenerational satanic ritual abuse or abuse at the age of six months. These cases were able to go forward because of changes in the statutes of limitation that permitted people to sue their parents, other relatives, teachers, doctors, and others if they claimed that they now remembered sexual abuse that had previously been repressed. The cases proceeded under the belief that when people are repeatedly brutalized, their memories can be completely repressed into the unconscious and later reliably recovered with hypnosis, dream interpretation, sodium amytal, or other therapeutic "memory work." In fact, no credible scientific support has been found for such claims.

After seeing the vast array of cases in which people sued their alleged abusers or brought them up on criminal charges in jurisdictions that allowed this, we began to see another sort of psychological and legal phenomenon. A large number of patients who came to believe as a result of questionable therapy that they had been extensively abused later concluded that their memories were false. Often having cut off their ties to family or even sought to destroy their families, many of these "retractors" sued their former therapists for planting the false memories. No tricky statute-of-limitation issues were involved here, as these were handled as traditional medical malpractice cases. The largest settlement to date was $10.6 million against a psychiatrist and major hospital in Chicago for a woman and her two young children who were led to believe falsely that they were victims of satanic ritual abuse and had developed multiple personalities. Even the young children were hospitalized for years under this dubious diagnosis, left to flounder with their incredible set of beliefs and false memories.

Then came the third-party lawsuits. Even when the "patients" had not retracted the beliefs, some family members sued the therapists for planting false memories in the mind of their adult child. The first substantial case to come to national attention involved the Ramona family. The daughter came to believe that her father had raped her for more than a decade, memories she acquired when she went into therapy as a sophomore in college. She sued her father, and he in turn sued the therapists who planted these beliefs. A jury in Napa, California, awarded him $500,000.

Then came the "Daddy-dead" cases. It was inconvenient when Daddy took the stand and convincingly denied any abuse, so some accusers waited until he died and then sued the estate. This left grieving widows and other heirs to defend against the abuse claims that might have dated back a quarter of a century. There were also the civil cases brought against corporations by those who claimed that the newly remembered abuse happened on their premises. They would claim that the alleged abuse took place in a McDonald's bathroom or on a Royal Caribbean cruise or in the high school art room. Even a well-funded corporation has a difficult time defending against supposedly repressed memories about events that purportedly happened thirty, forty, or fifty years ago.

Psychological studies have shown that it is virtually impossible to tell the difference between a real memory and one that is a product of imagination or some other process. Occasionally the memories could be shown to be false because they were biologically, geographically, or psychologically impossible. People remembered extensive abuse by a relative who was not living in the area at the time, or they remembered abuse that was supposed to have happened when they were one year old. The documented cases of false belief or memory illusion make it natural to wonder how it is that someone could come to believe that they had been sexually abused for years, and to even have very detailed memories, if in fact it never happened. Studies of memory distortion provide a clue. If there was anything good that came out of this decade of vitriolic controversy, it was a body of scientific research on memory that could leave a lasting positive contribution, at least in terms of its ability to help our understanding of the malleable nature of our memories.

The Science of Memory

For several decades, I and other psychological scientists have done research on memory distortion, specifically on showing how memories can be changed by things that we are told. Our memories are vulnerable to "post-event information": to details, ideas, and suggestions that come along after an event has happened. People integrate new materials into their memory, modifying what they believe they personally experienced. When people combine information gathered at the time of an actual experience with new information acquired later, they form a smooth and seamless memory and thereafter have great difficulty telling which facts came from which time.

More specifically, when people experience some actual event — say, a crime or an accident — they often later acquire new information about the event. This new information can contaminate the memory. This can happen when the person talks with other people, is exposed to media coverage about the event, or is asked leading questions. A simple question such as "How fast were the cars going when they smashed into each other?" has led experimental witnesses to an auto accident to estimate the speed of the cars as greater than did control witnesses who were asked a question like "How fast were the cars going when they hit each other?" Moreover, those asked the leading "smashed" question were more likely to claim to have seen broken glass, even though no glass had broken at all. Hundreds, perhaps thousands, of studies have revealed this kind of malleability of memory.

But post-event suggestion can do more than alter memory for a detail here and there from an actually experienced event; it can create entirely false memories. In the past few years, new research has shown just how far one can go in creating in the minds of people detailed memories of entire events that never occurred. Here are some examples.

As researchers, we wanted to find out if it was possible to deliberately plant a false memory. We set out by trying to convince subjects that they had been lost in a shopping mall at the age of five for an extended time and were ultimately rescued by an elderly person and reunited with the family. My colleague Jacquie Pickrell and I injected this pseudomemory into normal adults by enlisting the help of their mothers, fathers, and other older relatives, and by telling our subjects that the relatives had told us that these made-up experiences had happened. About a quarter of the subjects in our study fell sway to our suggestions and were led to believe, fully or partially, that they had been lost in this specific way.

Since the initial lost-in-the-mall study, numerous investigators have experimented with planting false memories, and many exceeded our initial levels of successful tampering. Taken together, these studies have taught us much about the memory distortion process. For example, one group of researchers at the University of British Columbia obtained facts about their subjects' childhoods from relatives and then attempted to elicit a false memory using guided imagery, context reinstatement, and mild social pressure, and by encouraging repeated attempts to recover the memory. The false memories the researchers tried to plant were events such as suffering a serious animal attack, a serious accident, or an injury by another child. They succeeded in creating a complete false memory in 26 percent of their subjects and a partial false memory in another 30 percent. Another research group from the University of Tennessee planted false memories of getting lost in a public place or being rescued by a lifeguard. With the help of techniques to stimulate the subject's imagination, they succeeded in 37 percent of their subjects. One false lifeguard rescue memory was quite detailed: "We went to the pool at the N the year we lived there. And my parents were lying by the pool, and I was in the shallow end with this kid I knew. And we started swimming toward the deep end, but we didn't get very far . . . and I remember he started to go under, and he grabbed me and pulled me under with him. And I remember being underwater and then hearing this big splash. He jumped in and just grabbed both of us at once and pulled us over to the side . . . And he was yelling at us."

Efforts to distinguish true from false memories revealed a few statistical differences. For example, the true memories were more emotionally intense than the false ones, and images in false memories were more likely to be viewed from the perspective of an observer, whereas images in true memories were more likely to be viewed from the first-person perspective. However, many of the differences between true and false memories are lessened or eliminated when the false memories are repeatedly rehearsed or retold. The statistical differences were never large enough to be able to take a single real-world memory report and reliably classify it as true or false.

The false memories of lifeguard rescues and other created events were helped along by the encouragement to use imagination. In other studies too, imagination has been a fruitful way to lead people to false memories. In one study, imagination succeeded in getting people to be more confident that as a child they had broken a window with their hand, and in another study imagination helped lead people to remember falsely that they kissed a plastic frog.

Imagination helps the false memory formation process in a number of ways. Some scientists have used the term "memory illusion" to refer to cases in which people have a false belief about the past that is experienced as a memory. In these cases, the person feels as if he or she is directly remembering some past event personally. By contrast, the term "false belief" applies to the case where the person has an incorrect belief about the past but doesn't feel as if this is being directly remembered. An insinuation or assertion that something happened can make someone believe that something happened: a false belief. But imagination supplies details that add substance to the belief. Rehearsal of these details can help to turn the false belief into a memory illusion.

One could argue that these studies bear little resemblance to the world of psychotherapy, which was so frequently implicated in the repressed memory legal cases. To address this, my Italian collaborator Giuliana Mazzoni and I attempted to create an experimental world that would be somewhat closer to the therapy experience. We began with the observation that dream interpretation is commonly used in psychotherapy. From ancient times, dreams have seemed mysterious and frequently prophetic. Modern bookstores are filled with books devoted solely or partly to the analysis of dream material, and some psychotherapists believe (as did Freud) that dream interpretation can lead to accurate knowledge about the patient's distant past. We wondered, however, whether dream work might be leading, not to an extraction of some buried but true past, but to the planting of a false past. In our first dream study, a large pool of undergraduates filled out a questionnaire to screen them about the likelihood of early childhood experiences happening to them. These included being lost for an extended period of time or feeling abandoned by their family before the age of three. We selected students who indicated that these experiences probably didn't happen to them.

Half of the subjects were selected to participate in what they thought was a completely different study, one that involved bringing a recent or recurring dream with them for analysis in a study of sleep and dreams. These subjects related their dreams to a trained clinician, an individual who happened to be a popular radio psychologist in Florence, Italy, where this first study was conducted. He told the subject about his extensive experience in dream interpretation and how it was that dreams reflected buried memories of the past. He talked to the subject about his or her ideas about the dream report and then offered his own interpretation. His analysis was always the same, no matter what the dream report: The dream indicated that the subject had some unhappiness related to a past experience that happened when the subject was very young and might not be remembered. His suggestions became even more specific: that the dream seemed to indicate that the subject had been lost for an extended time in a public place before age three, that the subject felt abandoned by his or her family, that the subject felt lonely and lost in an unfamiliar place. He stressed that these traumatic experiences could be buried in the subject's unconscious memory but were expressing themselves in the dream. The entire session with the clinician lasted about a half hour.

A couple of weeks later, the students returned to what they thought was the earlier study and once again filled out the screening questionnaire on their childhood experiences. Control subjects who had not been exposed to any dream interpretation responded pretty much as they had before. The majority of subjects whose dreams had been interpreted by the clinician became more confident that they had been lost in a public place before age three, that they had felt abandoned by their family, and that they had felt lonely and lost in an unfamiliar place. In a later study we tried to find out more about the phenomenological experience: Did subjects have a false belief or did they have a memory illusion? We found that about half the time our dream interpretation subjects ended up with a false belief and half the time with a memory illusion.

What is remarkable is that such large alterations of autobiography could be achieved so quickly. A half hour with the clinician is far less than the extensive and repeated dream interpretation that goes on in some psychotherapy that spans months or even years. Because many people enter therapy with the notion that dreams reveal real past events, and some therapists bolster this belief and freely suggest possible meanings, the potential for the personal past to become distorted in this way is very real. This is probably why a number of psychologists are now suggesting that dabbling in dream interpretation can be a dangerous activity. The psychologist Tana Dineen, in an essay entitled "Dangerous Dreaming," suggested that professionals should not pretend to know what dreams mean or that they reveal anything about the past.

These and other therapeutic interventions have been vigorously criticized in recent years because of the science-based fear that they encourage patients to concoct images of false events such as sexual abuse, to suppose that these images must be memories, and to act on them in destructive ways.

More Routes to Memory

People might think that avoiding certain types of psychotherapy where dream interpretation and imagination exercises are used renders them safe from unwanted intrusions into autobiography, but they should think again. There are other avenues by which fiction can creep into memory structures.

In fall 2000, I delivered a series of lectures in New Zealand and on one occasion offered up the prediction that we would see a rise in cases of demonic possession. I'm not sure that my audience took the news with the seriousness that they should have. But I knew a few things they didn't know. I knew about some recent findings on demonic possession, and I knew then that the famous film *The Exorcist* was soon to be re-released.

When I learned that *The Exorcist* would be re-released, I was prompted to look back at what happened in 1971 when William Blatty's book by that name was first published, followed two years later by the release of the film. Millions of people saw Linda Blair, as the twelve-year-old Regan, spewing vomit and waving a bloody crucifix. They saw various priests perform an exorcism on her. What followed were reports of fainting and vomiting during the film, mass hysteria in the form of symptoms of vomiting, fainting, and trembling, and a mini-epidemic of supposed possession. People sought exorcisms in record numbers. In the words of the sociologist Michael Cuneo, "Thousands of households across America seemed to become infested all of a sudden with demonic presences, and Catholic rectories were besieged with calls from people seeking exorcisms for themselves, for their loved ones, and sometimes even for their pets." Cuneo did an interview with Father Tom Bermingham, who had played a minor role in the film and received screen credit as a technical adviser: "When the movie came out, I found myself on the hot seat. People saw my face and my name on the screen, and they assumed I was the answer to their problems. For quite a while dozens of people were trying to contact me every week. And they weren't all Catholics. Some were Jewish, some Protestant, some agnostic, and they all believed that they themselves or someone close to them might be demonically possessed. They were truly desperate people."

What was going on? In giving visual form to a phenomenon, *The*

Exorcist and other films and stories like it convinced people that possession by the devil was plausible, that possession was more than a possibility. Some people were led even further — to actual belief and symptoms. How could this happen? Can it happen only to people who already think that demonic possession is plausible?

Based on a series of studies conducted with Giuliana Mazzoni of Seton Hall University and Irving Kirsch of the University of Connecticut, we understand some of the process. In the first of these studies, subjects first rated the plausibility of a number of events and gave information about their childhood experiences, including the event of witnessing demonic possession as a child. Later, some subjects read several short articles that described demonic possession, suggesting that it was more common than previously thought, and described typical possession experiences. Subjects also took a "fear profile," in which their particular fears were analyzed; whatever their responses on the profile, they were given the false feedback that witnessing a possession during childhood probably caused those fears. In the final phase of the study, subjects once again rated the plausibility of life events and gave information about their own childhood experiences. Relative to control subjects, those who were exposed to the possession manipulation increased the plausibility of witnessing possession but also made a number of individual claims that it had happened to them.

In follow-up studies, we found that the stories alone could produce some influence and that stories that were set in contemporary culture were more effective than those set in some remote time and culture. Taken together, the studies show that reading a few stories and hearing about another individual's experience can increase plausibility and make you more confident that something, even something implausible, happened to you. A major point worth emphasizing is that the suggestive material in the study worked not only with people who began with the belief that demonic possession was plausible but also with those who began with the belief that it was rather implausible. The studies constitute the beginning of a recipe for making the implausible seem plausible and sending someone down the road to developing a full-blown false memory.

Back to the prediction I made to that New Zealand audience that demonic possession would soon be on the rise. On September 22,

2000, *The Exorcist* was re-released with eleven added minutes of original footage. On Halloween, there was a broadcast of *Possessed,* a TV docudrama about a purported exorcism in a mental hospital. By the end of November, the *New York Times* was reporting that new exorcism teams had been assembled in response to increased public demand. In New Zealand, I'm receiving a lot more respect. This is an example of how the mass media can mythologize reality. It can show us something we have never seen and might never even have imagined otherwise. In this way it gains a pervasive influence over our consciousness in its power to fashion reality for us.

No Escape

Lest you think you might stop watching films and television programs, stop reading magazine stories, and find refuge in the advertisements, that might not help. Even this material has the power to tamper with autobiography. Kathryn Braun, Rhiannon Ellis, and I designed a series of studies in which we used advertising copy to try to plant memories. In one study, subjects filled out questionnaires and answered questions about a trip to Disneyland. One group read and evaluated a fake Disneyland ad featuring Bugs Bunny and describing how they met and shook hands with the character. About 16 percent of the people who evaluated the fake Bugs ad later said that they had personally met Bugs Bunny when they visited Disneyland. Later studies showed that with multiple exposures to phony Disney ads involving Bugs, the percentages rose to roughly 30 percent. The problem is that Bugs is a Warner Brothers character not to be found at Disneyland. Despite the impossibility of this false memory, significant numbers were influenced to remember meeting him and ultimately also became more likely to relate Bugs Bunny to other Disney concepts such as Mickey Mouse or the Magic Castle.

We are not suggesting that advertisers are actually planting false memories deliberately. After all, you would not in reality see an ad for Disney that featured Bugs Bunny. But you might see one featuring a handshake with Mickey Mouse, and this would increase confidence that the viewer personally experienced such a handshake. The memory might be true for some people, but it is certainly not true for all. In this way, the advertisements may actually be tampering with our childhood memories in ways that we're not even aware of.

What Does It All Mean?

Medieval and modern philosophical accounts of human cognition stressed the role of imagination. The eighteenth-century philosopher Immanuel Kant talked about imagination as the faculty for putting together various mental representations such as sense percepts, images, and concepts. This integrative activity bears a great resemblance to what memory actually is and does. We see a film, it feeds into our dreams, it seeps into our memories. Our job as researchers in this area is to understand how it is that pieces of experience are combined to produce what we experience as "memory." All memory involves reconstruction. We put together pieces of episodes that are not well connected, and we continually make judgments about whether a particular piece belongs in the memory or not. One expects to see a shuffling of pieces with a process that works like this.

As scientists work toward understanding how false autobiographical memories come to be, we'll understand ourselves better, but we will also have a better handle on how such errors might be prevented.

What shall we do with all we have learned about the malleable nature of memory? We might start by recognizing that a reconstructed memory that is partly fact and partly fiction might be good enough for many facets of life but inadequate for legal purposes, where very precise memory often matters. It matters whether the light was red or green, whether the driver of the getaway car had straight hair or curly. It matters whether that face is the face of the person who committed the murder. Keep in mind that some two hundred people per day in the United States become criminal defendants after being identified from lineups or photo spreads. The growing number of wrongfully convicted individuals who have been exonerated by DNA evidence has given the world a real appreciation of the problem of faulty eyewitness memory, which is the major cause of wrongful conviction. Faced with the horror of these recent cases, investigations by the U.S. Department of Justice, the Canadian government, and an Illinois Commission on Capital

Punishment have resulted in strong and specific recommendations designed to reduce the prevalence of wrongful convictions. Many of the recommendations reflect a heightened appreciation of the malleable nature of memory.

The U.S. Department of Justice released a 1996 report after analyzing twenty-eight cases of DNA exonerations and concluding that 80 percent of these innocent people had been convicted because of faulty eyewitness memory. The Justice Department then assembled a committee that came up with a set of guidelines for law enforcement. *Eyewitness Evidence: A Guide for Law Enforcement* offers a set of national guidelines for the collection and preservation of eyewitness evidence. The guide includes recommendations such as asking open-ended questions, not interrupting eyewitnesses' responses, and avoiding leading questions. It includes guidelines specifying how lineups should be constructed (for example, including only one true suspect per lineup and including the proper number of "fillers"). The publication, which makes use of psychological findings and explicitly acknowledges that these findings offer the legal system a valuable body of empirical knowledge, is not a legal mandate but rather a document that hopes to promote sound professional practice. Nevertheless, it is apparently having an influence on actual practice, and those who deviate significantly from it are often forced under cross-examination to say why.

The Canadians were also rocked by cases of wrongful conviction, prominent among them the case of Thomas Sophonow. He had been wrongfully convicted of murdering a young waitress who worked in a doughnut shop and spent nearly four years in prison. An official inquiry was established to investigate what went wrong, to determine just compensation for Mr. Sophonow, and to make recommendations about future cases. Commissioner Peter Cory was eloquent in his description of the suffering of this one falsely accused man: "What has he suffered? . . . He is psychologically scarred for life. He will always suffer from the core symptoms of post-traumatic stress disorder. As well, he will always suffer from paranoia, depression, and the obsessive desire to clear his name. His reputation as a murderer has affected him in every aspect of his life, from work to family relations. The community in which he lived believed him to be the murderer of a young woman, and that the crime had intimations of sexual assault. The damage to his reputation could not be greater . . . His reputation as a murderer will follow him wherever he goes. There will always be someone to whisper a false innuendo . . . In the mind of Thomas Sophonow, he will always believe that people are talking about him and his implication in the murder." Commissioner Cory awarded $1.75 million in nonpecuniary damages with a total award exceeding $2.5 million. To minimize future miscarriages of justice, the inquiry report on the Sophonow case calls for specific procedural changes in activities such as lineups, as well as more general guidance such as encouraging judges to emphasize to juries the frailties of memory, to recount the tragedies of wrongful convictions, and to readily admit expert testimony on the subject of memory.

A final example comes from Illinois. In March 2000, shortly after Governor Ryan declared a moratorium on executions in the state, he appointed a commission to determine what reforms, if any, would make the state's capital punishment system fair and just. These activities were prompted in part by the release of thirteen men from death row during the preceding decade. Many of these had been exonerated by DNA evidence. Steven Smith had been sentenced to death on the dubious testimony of a single eyewitness. Anthony Porter had been sentenced to death because of two eyewitnesses. They later recanted, and another man subsequently confessed and is now in prison. The commission made eighty-five recommendations, many of which flowed from a concern about faulty memory. They include training in the science of memory for police, prosecutors, and defense lawyers and the development of jury instructions to educate the jurors about factors that can affect eyewitness memory.

The Need for Education

These studies all recognize the need for education in order to integrate psychological science into law and courtroom practice. Judges, jurors, attorneys, and police will almost certainly be helped by an increased understanding of human memory. At a minimum, it is important to fully appreciate that false memory reports can look like true ones and that without independent corroboration it is virtually impossible to tell whether a particular report is the product of true memory or the product of imagination, suggestion, or some other process. Judges and juries sometimes think that they can tell the difference, but they are actually responding to the confidence, the detail, and the emotion with which a memory report is delivered. Unfortunately, these characteristics do not necessarily correspond with reliability.

How shall we educate people about the science of memory? It's not quite as simple as the late Carl Sagan's exhortation to teach more about the fundamentals of science in school. Education helps, but it has not protected people from embracing unsubstantiated beliefs such as paranormal phenomena, alien abduction, extraterrestrial visitors, telepathy, or communication with the dead. One effort to reduce these types of beliefs that had some early success involved getting students to participate actively in studies that reveal how such claims can be faked. In the current domain, we might consider not just asserting particular truths about memory but actually showing how studies have been done and what findings have been achieved.

Judges and jurors need to appreciate a point that can't be stressed enough: True memories cannot be distinguished from false without corroboration. Occasionally mental health professionals enter legal cases as expert witnesses and claim that they can tell that a "victim" is telling an accurate story. These purported experts frequently are there to bolster accusations that might otherwise seem strange. Beware of them. As Supreme Court Justice Stephen Breyer wrote two years ago in *Issues* ("Science in the Courtroom," Summer 2000), "Most judges lack the scientific training that might facilitate the evaluation of scientific claims or the evaluation of expert witnesses who make such claims." Education can help enhance the appreciation of good scientific information about memory as well as giving judges and jurors the confidence to reject pseudoscientific claims about memory.

Scientific knowledge about memory could be imparted in numerous venues: seminars for judges, law school classes for prospective attorneys, training for police, jury instructions, or expert testimony for jurors. This preliminary and tentative list could be expanded and refined through a cooperative effort by legal and scientific experts to develop a workable program for action. The American Judicature Society, an educational and research organization, recently proposed the creation of an "innocence commission," which would study why the legal system fails in ways that are reminiscent of what the National Transportation Safety Board does when planes crash. A National Memory Safety Board has a nice ring to me.

And What About the Priests?

The past decade produced innumerable casualties associated with claims of repressed or dissociated memories. As we cope with the recent revelations about abuse by Catholic priests, is there a lesson to be learned? As Dorothy Rabinowitz of the *Wall Street Journal* noted, these new revelations bring home the contrast between bogus charges and credible ones. Many victims of priest abuse had long histories of molestation, repeated over and over, with contemporaneous complaints that were recorded, even if they were hidden from the public. Other victims knew all along about their abuse, even if they never talked about it. There are few claims of abuse at age six months, or claims of impregnation at the age of six, or claims of abuse in intergenerational satanic rituals adorning these reports. But just as there was real sex abuse before the bogus repressed memory claims emerged, so there will be a mix of real and false accusations against priests, especially because there is the possibility of cash awards for damages. Not only will deliberate frauds emerge, but there will be "victims" who will, through suggestive therapy or media coverage, come to believe that they have been abused by priests when they have not. Publicizing the names of every single priest who might ever have been accused and firing priests simply on the strength of accusations is unfair and unjustified.

After the thousands of criminal charges and lawsuits against alleged abusers, we can expect to see retractors who sue their therapists and falsely accused individuals who sue their accusers and those who helped them develop the accusations. Large sums will be paid not only to those who bring the accusations but also later to those who claim they were falsely accused. It will not be a pretty sight. Apart from the lawsuits, there is the human damage. We've seen the names of the accused prominently featured on the front pages and airwaves before there is any sort of investigation. Cardinal Roger Mahony of Los Angeles saw his name in the headlines because of a single accusation by a fifty-one-year-old woman who had been previously diagnosed with schizophrenia. The *Los Angeles Times* drew parallels between the case of Mahony and that of the late Cardinal Joseph Bernardin. In a civil lawsuit filed against Bernardin in 1993, Steven Cook, a thirty-four-year-old seminarian, charged — on the basis of "recovered memory" induced through hypnosis — that Bernardin had sexually abused him seventeen years earlier. He sued for $10 million. The cardinal was "startled and devastated" by the accusation. I was an expert witness in that case and saw close up how dubious the memory recovery was, including the pieces brought out by a massage therapist. Eventually Cook retracted the accusation and apologized. Bernardin forgave him. Although he experienced a newfound sympathy for those falsely accused, the cardinal demonstrated a strengthened resolve to reach out to genuine victims of sexual abuse. Bernardin died of pancreatic cancer in 1996, not long after his accuser had died of AIDS. In the book that he completed thirteen days before his death, he singled out his cancer and the false accusation as the "major events" of his life. Although he lived a busy life marked by enough distinguished accomplishments and good works to fill several obituaries, virtually every obituary written after his death found space to mention the allegation of sexual abuse.

The parallel accusation against Mahony was front-page news for days. His accuser claimed that one day, thirty-two years earlier when she was in high school, she passed out near the band room, and when she awoke her pants were off and she saw Mahony's face. The police investigated the charge and found it groundless. A careful reader could have seen this reported in the press later the same month. What should we expect to find when his obituary is written?

The example should be a warning of the importance of keeping in mind just who we are. We're a nation that developed a legal system based first and foremost on due process. Of course we believe that it is important to punish evildoers, but we also have to balance that with the need to protect the innocent. If we ever lose that core element of our justice system, we will lose something that will ultimately cause us a grief far greater than we have ever known. As the church scandal gains momentum, perhaps we should have a commission of respected leaders whose role it is to keep the accusations in perspective and to convince everyone to withhold judgment until the facts are in.

If knowledge about human memory were to help reduce even

slightly the likelihood of wrongful accusations, the benefit for the accused and his or her extended family would be obvious. Society would also be better off, because while the wrong person is jailed, the real one is sometimes out and about committing further crimes.

But knowledge about human memory can help many others. When patients in therapy are being treated under the unsubstantiated belief that they have repressed memories of childhood trauma and that those memories must be excavated, this may not be doing the patients any good. If patients are diverted from the true cause of their problems and from seeking professional help that would actually make them better, they are harmed.

The mental health profession has also suffered from a proliferation of dubious beliefs about memory. The ridicule of a subgroup with questionable memory beliefs drags down the reputation of the entire profession. And finally, there is one last group that is harmed by a system that accepts every single claim of victimization no matter how dubious. That system dilutes and trivializes the experiences of the genuine victims and increases their suffering.

CHARLES C. MANN

Homeland Insecurity

FROM *The Atlantic Monthly*

- To stop the rampant theft of expensive cars, manufacturers in the 1990s began to make ignitions very difficult to hot-wire. This reduced the likelihood that cars would be stolen from parking lots — but apparently contributed to the sudden appearance of a new and more dangerous crime, carjacking.

- After a vote against management, Vivendi Universal announced earlier this year that its electronic shareholder-voting system, which it had adopted to tabulate votes efficiently and securely, had been broken into by hackers. Because the new system eliminated the old paper ballots, recounting the votes — or even independently verifying that the attack had occurred — was impossible.

- To help merchants verify and protect the identity of their customers, marketing firms and financial institutions have created large computerized databases of personal information: Social Security numbers, credit card numbers, telephone numbers, home addresses, and the like. With these databases being increasingly interconnected by means of the Internet, they have become irresistible targets for criminals. From 1995 to 2000 the incidence of identity theft tripled.

AS WAS OFTEN THE CASE, Bruce Schneier was thinking about a really terrible idea. We were driving around the suburban industrial wasteland south of San Francisco, on our way to a corporate presentation, while Schneier looked for something to eat not purveyed by a chain restaurant. This was important to Schneier, who in addition to being America's best-known ex-cryptographer is a food writer for an alternative newspaper in Minneapolis, where he lives. Initially he had been sure that in the crazy ethnic salad of Silicon Valley it would be impossible not to find someplace of culinary interest — a Libyan burger stop, a Hmong bagelry, a Szechuan taco stand. But as the rented car swept toward the vast, amoeboid office complex that was our destination, his faith slowly crumbled. Bowing to reality, he parked in front of a nondescript sandwich shop, disappointment evident on his face.

Schneier is a slight, busy man with a dark, full, closely cropped beard. Until a few years ago, he was best known as a prominent creator of codes and ciphers; his book *Applied Cryptography* (1993) is a classic in the field. But despite his success he virtually abandoned cryptography in 1999 and co-founded a company named Counterpane Internet Security. Counterpane has spent considerable sums on advanced engineering, but at heart the company is dedicated to bringing one of the oldest forms of policing — the cop on the beat — to the digital realm. Aided by high-tech sensors, human guards at Counterpane patrol computer networks, helping corporations and governments to keep their secrets secret. In a world that is both ever more interconnected and full of malice, this is a task of considerable difficulty and great importance. It is also what Schneier long believed cryptography would do — which brings us back to his terrible idea.

"Pornography!" he exclaimed. If the rise of the Internet has shown anything, it is that huge numbers of middle-class, middle-management types like to look at dirty pictures on computer screens. A good way to steal the corporate or government secrets these middle managers are privy to, Schneier said, would be to set up a pornographic Web site. The Web site would be free, but visitors would have to register to download the naughty bits. Registration would involve creating a password — and here Schneier's deep-set blue eyes widened mischievously.

People have trouble with passwords. The idea is to have a random string of letters, numbers, and symbols that is easy to remember. Alas, random strings are by their nature hard to remember, so people use bad but easy-to-remember passwords, such as "hello" and "password." (A survey last year of 1,200 British office workers found that almost half chose their own name, the name of a pet, or that of a family member as a password; others based their passwords on the names Darth Vader and Homer Simpson.) Moreover, computer users can't keep different passwords straight, so they use the same bad passwords for all their accounts.

Many of his corporate porn surfers, Schneier predicted, would use for the dirty Web site the same password they used at work. Not only that, many users would surf to the porn site on the fast Internet connection at the office. The operators of Schneier's nefarious site would thus learn that, say, "Joesmith," who accessed the Web site from Anybusiness.com, used the password "JoeS." By trying to log on at Anybusiness.com as "Joesmith," they could learn whether "JoeS" was also the password into Joesmith's corporate account. Often it would be.

"In six months you'd be able to break into Fortune 500 companies and government agencies all over the world," Schneier said, chewing his nondescript meal. "It would work! It would work — that's the awful thing."

During the 1990s Schneier was a field marshal in the disheveled army of computer geeks, mathematicians, civil liberties activists, and libertarian wackos that — in a series of bitter lawsuits that came to be known as the Crypto Wars — asserted the right of the U.S. citizenry to use the cryptographic equivalent of kryptonite: ciphers so powerful they cannot be broken by any government, no matter how long and hard it tries. Like his fellows, he believed that "strong crypto," as these ciphers are known, would forever guarantee the privacy and security of information — something that in the Information Age would be vital to people's lives. "It is insufficient to protect ourselves with laws," he wrote in *Applied Cryptography*. "We need to protect ourselves with mathematics."

Schneier's side won the battle as the nineties came to a close. But by that time he had realized that he was fighting the wrong war. Crypto was not enough to guarantee privacy and security. Failures occurred all the time — which was what Schneier's terrible idea demonstrated. No matter what kind of technological safeguards an organization uses, its secrets will never be safe while its employees are sending their passwords, however unwittingly, to pornographers — or to anyone else outside the organization.

The Parable of the Dirty Web Site illustrates part of what became the thesis of Schneier's most recent book, *Secrets and Lies* (2000): The way people think about security, especially security on computer networks, is almost always wrong. All too often planners seek technological cure-alls, when such security measures at best limit risks to acceptable levels. In particular, the consequences of going wrong — and all these systems go wrong sometimes — are rarely considered. For these reasons, Schneier believes that most of the security measures envisioned after September 11 will be ineffective and that some will make Americans *less* safe.

It is now a year since the World Trade Center was destroyed. Legislators, the law enforcement community, and the Bush administration are embroiled in an essential debate over the measures necessary to prevent future attacks. To armor-plate the nation's security, they increasingly look to the most powerful technology available: retina, iris, and fingerprint scanners; "smart" driver's licenses and visas that incorporate anti-counterfeiting chips; digital surveillance of public places with face recognition software; huge centralized databases that use data mining routines to sniff out hidden terrorists. Some of these measures have already been mandated by Congress, and others are in the pipeline. State and local agencies around the nation are adopting their own schemes. More mandates and more schemes will surely follow.

Schneier is hardly against technology — he's the sort of person who immediately cases public areas for outlets to recharge the batteries in his laptop, phone, and other electronic prostheses. "But if you think technology can solve your security problems," he says, "then you don't understand the problems and you don't understand the technology." Indeed, he regards the national push for a high-tech salve for security anxieties as a reprise of his own early and erroneous beliefs about the transforming power of strong crypto. The new technologies have enormous capacities, but their advocates have not realized that the most critical aspect of a security measure is not how well it works but how well it fails.

The Crypto Wars

If mathematicians from the 1970s were suddenly transported through time to the present, they would be happily surprised by developments such as the proofs to Kepler's conjecture (proposed in 1611, confirmed in 1998) and to Fermat's last theorem (1637, 1994). But they would be absolutely astonished by the RSA Conference, the world's biggest trade show for cryptographers. Sponsored by the cryptography firm RSA Security, the conferences are attended by as many as 10,000 cryptographers, computer scientists, network managers, and digital security professionals. What would amaze past mathematicians is not just the number of conferences but that they exist at all.

Cryptology is a specialized branch of mathematics with some computer science thrown in. As recently as the 1970s there were no cryptology courses in university mathematics or computer science departments; nor were there crypto textbooks, crypto journals, or crypto software. There was no private crypto industry, let alone venture-capitalized crypto start-ups giving away key rings at trade shows (*crypto key* rings — technohumor). Cryptography, the practice of cryptology, was the province of a tiny cadre of obsessed amateurs, the National Security Agency, and the NSA's counterparts abroad. Now it is a multibillion-dollar field with applications in almost every commercial arena.

As one of the people who helped to bring this change about, Schneier is always invited to speak at RSA conferences. Every time, the room is too small, and overflow crowds, eager to hear their favorite guru, force the session into a larger venue, which is what happened when I saw him speak at an RSA conference in San Francisco's Moscone Center last year. There was applause from the hundreds of seated cryptophiles when Schneier mounted the stage, and more applause from the throng standing in the aisles and exits when he apologized for the lack of seating capacity. He was there to talk about the state of computer security, he said. It was as bad as ever, maybe getting worse.

In the past, security officers were usually terse ex-military types who wore holsters and brush cuts. But as computers have become both attackers' chief targets and their chief weapons, a new generation of security professionals has emerged, drawn from the ranks of engineering and computer science. Many of the new guys look like people the old guard would have wanted to arrest, and Schneier is no exception. Although he is a co-founder of a successful company, he sometimes wears scuffed black shoes and pants with a wavering press line; he gathers his thinning hair into a straggly ponytail. Ties, for the most part, are not an issue. Schneier's style marks him as a true nerd — someone who knows the potential, both good and bad, of technology, which in our technocentric era is an asset.

Schneier was raised in Brooklyn. He got a B.S. in physics from the University of Rochester in 1985 and an M.S. in computer science from American University two years later. Until 1991 he worked for the Department of Defense, where he did things he won't discuss. Lots of kids are intrigued by codes and ciphers, but Schneier was surely one of the few to ask his father, a lawyer and a judge, to write secret messages for him to analyze. On his first visit to a voting booth, with his mother, he tried to figure out how she could cheat and vote twice. He didn't actually want her to vote twice — he just wanted, as he says, to "game the system."

Unsurprisingly, someone so interested in figuring out the secrets of manipulating the system fell in love with the systems for manipulating secrets. Schneier's childhood years, as it happened, were a good time to become intrigued by cryptography — the best time in history, in fact. In 1976 two researchers at Stanford University invented an entirely new type of encryption, public key encryption, which abruptly woke up the entire field.

Public key encryption is complicated in detail but simple in outline. All ciphers employ mathematical procedures called algorithms to transform messages from their original form into an unreadable jumble. (Cryptographers work with ciphers and not codes, which are spy movie–style lists of prearranged substitutes for letters, words, or phrases — "meet at the theater" for "attack at nightfall.") Most ciphers use secret keys: mathematical values that plug into the algorithm. Breaking a cipher means figuring out the key. In a kind of mathematical sleight of hand, public key encryption encodes messages with keys that can be published openly and decodes them with different keys that stay secret and are effectively impossible to break, using today's technology.

The best-known public key algorithm is the RSA algorithm, whose name comes from the initials of the three mathematicians who invented it. RSA keys are created by manipulating big prime numbers. If the private decoding RSA key is properly chosen, guessing it necessarily involves factoring a very large number into its constituent primes, something for which no mathematician has ever devised an adequate shortcut. Even if demented government agents spent a trillion dollars on custom factoring computers, Schneier has estimated, the sun would likely go nova before they cracked a message enciphered with a public key of sufficient length.

Schneier and other technophiles grasped early how important computer networks would become to daily life. They also understood that those networks were dreadfully insecure. Strong crypto, in their view, was an answer of almost magical efficacy. Even federal officials believed that strong crypto would Change Everything Forever — except they thought the change would be for the worse. Strong encryption "jeopardizes the public safety and national security of this country," Louis Freeh, then the director of the (famously computer-challenged) Federal Bureau of Investigation, told Congress in 1995. "Drug cartels, terrorists, and kidnappers will use telephones and other communications media with impunity, knowing that their conversations are immune" from wiretaps.

The Crypto Wars erupted in 1991, when Washington attempted to limit the spread of strong crypto. Schneier testified before Congress against restrictions on encryption, campaigned for crypto freedom on the Internet, co-wrote an influential report on the technical snarls awaiting federal plans to control cryptographic protocols, and rallied 75,000 crypto fans to the cause in his free monthly e-mail newsletter, *Crypto-Gram* (www.counterpane.com/crypto-gram.html). Most important, he wrote *Applied Cryptography*, the first-ever comprehensive guide to the practice of cryptology.

Washington lost the wars in 1999, when an appellate court ruled that restrictions on cryptography were illegal, because crypto algorithms were a form of speech and thus covered by the First Amendment. After the ruling, the FBI and the NSA more or less surrendered. In the sudden silence, the dazed combatants surveyed the battleground. Crypto had become widely available, and it had indeed fallen into unsavory hands. But the results were different from what either side had expected.

As the crypto aficionados had envisioned, software companies inserted crypto into their products. On the "Tools" menu in Microsoft Outlook, for example, "encrypt" is an option. And encryption became big business as part of the infrastructure for e-commerce — it is the little padlock that appears in the corner of Net surfers' browsers when they buy books at Amazon.com, signifying that credit card numbers are being enciphered. But encryption is rarely used by the citizenry it was supposed to protect and empower. Cryptophiles, Schneier among them, had been so enraptured by the possibilities of uncrackable ciphers that they forgot they were living in a world in which people can't program VCRs. Inescapably, an encrypted message is harder to send than an unencrypted one, if only because of the effort involved in using all the extra software. So few people use encryption software that most companies have stopped selling it to individuals.

Among the few who do use crypto are human rights activists living under dictatorships. But, just as the FBI feared, terrorists, child pornographers, and the Mafia use it too. Yet crypto has not protected any of them. As an example, Schneier points to the case of Nicodemo Scarfo, who the FBI believed was being groomed to take over a gambling operation in New Jersey. Agents surreptitiously searched his office in 1999 and discovered that he was that rarity, a gangster nerd. On his computer was the long-awaited nightmare for law enforcement: a crucial document scrambled by strong encryption software. Rather than sit by, the FBI installed a "keystroke logger" on Scarfo's machine. The logger recorded the decrypting key — or, more precisely, the passphrase Scarfo used to generate that key — as he typed it in, and gained access to his incriminating files. Scarfo pleaded guilty to charges of running an illegal gambling business on February 28 of this year.

Schneier was not surprised by this demonstration of the impotence of cryptography. Just after the Crypto Wars ended, he had begun writing a follow-up to *Applied Cryptography*. But this time Schneier, a fluent writer, was blocked — he couldn't make himself extol strong crypto as a security panacea. As Schneier put it in *Secrets and Lies,* the very different book he eventually did write, he had been portraying cryptography — in his speeches, in his congressional testimony, in *Applied Cryptography* — as "a kind of magic security dust that [people] could sprinkle over their software and make it secure." It was not. Nothing could be. Humiliatingly, Schneier discovered that, as a friend wrote him, "the world was full of bad security systems designed by people who read *Applied Cryptography.*"

In retrospect he says, "Crypto solved the wrong problem." Ciphers scramble messages and documents, preventing them from being read while, say, they are transmitted on the Internet. But the strongest crypto is gossamer protection if malevolent people have access to the computers on the other end. Encrypting transactions on the Internet, the Purdue computer scientist Eugene Spafford has remarked, "is the equivalent of arranging an armored car to deliver credit card information from someone living in a cardboard box to someone living on a park bench."

To effectively seize control of Scarfo's computer, FBI agents had to break into his office and physically alter his machine. Such black-bag jobs are ever less necessary, because the rise of networks and the Internet means that computers can be controlled remotely, without their operators' knowledge. Huge computer databases may be useful, but they also become tempting targets for criminals and terrorists. So do home computers, even if they are connected only intermittently to the Web. Hackers look for vulnerable machines, using software that scans thousands of Net connections at once. This vulnerability, Schneier came to think, is the real security issue.

With this realization he closed Counterpane Systems, his five-person crypto-consulting company in Chicago, in 1999. He revamped it and reopened immediately in Silicon Valley with a new name, Counterpane Internet Security, and a new idea — one that relied on old-fashioned methods. Counterpane would still keep data secret. But the lessons of the Crypto Wars had given Schneier a different vision of how to do that — a vision that has considerable relevance for a nation attempting to prevent terrorist crimes.

Where Schneier had sought one overarching technical fix, hard experience had taught him the quest was illusory. Indeed, yielding to the American penchant for all-in-one high-tech solutions can make us *less* safe — especially when it leads to enormous databases full of confidential information. Secrecy is important, of course, but it is also a trap. The more secrets necessary to a security system, the more vulnerable it becomes.

To forestall attacks, security systems need to be small in scale, redundant, and compartmentalized. Rather than large, sweeping programs, they should be carefully crafted mosaics, each piece aimed at a specific weakness. The federal government and the airlines are spending millions of dollars, Schneier points out, on systems that screen every passenger to keep knives and weapons out of planes. But what matters most is keeping dangerous passengers out of airline cockpits, which can be accomplished by reinforcing the door. Similarly, it is seldom necessary to gather large amounts of additional information, because in modern societies people leave wide audit trails. The problem is sifting through the already existing mountain of data. Calls for heavy monitoring and record-keeping are thus usually a mistake. ("Broad surveillance is a mark of bad security," Schneier wrote in a recent *Crypto-Gram*.)

To halt attacks once they start, security measures must avoid being subject to single points of failure. Computer networks are particularly vulnerable: once hackers bypass the firewall, the whole system is often open for exploitation. Because every security measure in every system can be broken or gotten around, failure must be incorporated into the design. No single failure should compromise the normal functioning of the entire system or, worse, add to the gravity of the initial breach. Finally, and most important, decisions need to be made by people at close range — and the responsibility needs to be given explicitly to people, not computers.

Unfortunately, there is little evidence that these principles are playing any role in the debate in the administration, Congress, and the media about how to protect the nation. Indeed, in the argument over policy and principle, almost no one seems to be paying attention to the practicalities of security — a lapse that Schneier, like other security professionals, finds as incomprehensible as it is dangerous.

Stealing Your Thumb

A couple of months after September 11, I flew from Seattle to Los Angeles to meet Schneier. As I was checking in at Sea-Tac Airport, someone ran through the metal detector and disappeared onto the little subway that runs among the terminals. Although the authorities quickly identified the miscreant, a concession stand worker, they still had to empty all the terminals and rescreen everyone in the airport, including passengers who had already boarded planes. Masses of unhappy passengers stretched back hundreds of feet from the checkpoints. Planes by the dozen sat waiting at the gates. I called Schneier on a cell phone to report my delay. I had to shout over the noise of all the other people on their cell phones making similar calls. "What a mess," Schneier said. "The problem with airport security, you know, is that it fails badly."

For a moment I couldn't make sense of this gnomic utterance.

Then I realized he meant that when something goes wrong with security, the system should recover well. In Seattle a single slip-up
shut down the entire airport, which delayed flights across the nation. Sea-Tac, Schneier told me on the phone, had no adequate
way to contain the damage from a breakdown — such as a button
installed near the x-ray machines to stop the subway, so that idiots
who bolt from checkpoints cannot disappear into another terminal. The shutdown would inconvenience subway riders, but not as
much as being forced to go through security again after a wait of
several hours. An even better idea would be to place the x-ray machines at the departure gates, as some are in Europe, in order to
scan each group of passengers closely and minimize inconvenience
to the whole airport if a risk is detected — or if a machine or a
guard fails.

Schneier was in Los Angeles for two reasons. He was to speak to
ICANN, the Internet Corporation for Assigned Names and Numbers, which controls the "domain name system" of Internet addresses. It is Schneier's belief that attacks on the address database
are the best means of taking down the Internet. He also wanted to
review Ginza Sushi-Ko, perhaps the nation's most exclusive restaurant, for the food column he writes with his wife, Karen Cooper.

Minutes after my delayed arrival, Schneier had with characteristic celerity packed himself and me into a taxi. The restaurant was in
a shopping mall in Beverly Hills that was disguised to look like a
collection of nineteenth-century Italian villas. By the time Schneier
strode into the tiny lobby, he had picked up the thread of our airport discussion. Failing badly, he told me, was something he had
been forced to spend time thinking about.

In his technophilic exuberance, he had been seduced by the
promise of public key encryption. But ultimately Schneier observed that even strong crypto fails badly. When something by-
passes it, as the keystroke logger did with Nicodemo Scarfo's encryption, it provides no protection at all. The moral, Schneier
came to believe, is that security measures are characterized less by
their manner of success than by their manner of failure. All security systems eventually miscarry. But when this happens to the good
ones, they stretch and sag before breaking, each component failure leaving the whole as unaffected as possible. Engineers call such
failure-tolerant systems "ductile." One way to capture much of what

Schneier told me is to say that he believes that when possible, security schemes should be designed to maximize ductility, whereas they often maximize strength.

Since September 11, the government has been calling for a new security infrastructure — one that employs advanced technology to protect the citizenry and track down malefactors. Already the USA PATRIOT Act, which Congress passed in October, mandates the establishment of a "cross-agency, cross-platform electronic system . . . to confirm the identity" of visa applicants, along with a "highly secure network" for financial-crime data and "secure information sharing systems" to link other, previously separate databases. Pending legislation demands that the attorney general employ "technology including, but not limited to, electronic fingerprinting, face recognition, and retinal scan technology." The proposed Department of Homeland Security is intended to oversee a "national research and development enterprise for homeland security comparable in emphasis and scope to that which has supported the national security community for more than fifty years" — a domestic version of the high-tech R&D juggernaut that produced stealth bombers, smart weapons, and antimissile defense.

Iris, retina, and fingerprint scanners; hand-geometry assayers; remote video network surveillance; face recognition software; smart cards with custom identification chips; decompressive baggage checkers that vacuum-extract minute chemical samples from inside suitcases; tiny radio implants beneath the skin that continually broadcast people's identification codes; pulsed fast-neutron analysis of shipping containers ("so precise," according to one manufacturer, "it can determine within inches the location of the concealed target"); a vast national network of interconnected databases — the list goes on and on. In the first five months after the terrorist attacks, the Pentagon liaison office that works with technology companies received more than 12,000 proposals for high-tech security measures. Credit card companies expertly manage credit risks with advanced information-sorting algorithms, Larry Ellison, the head of Oracle, the world's biggest database firm, told the *New York Times* in April; "We should be managing security risks in exactly the same way." To "win the war on terrorism," a former deputy undersecretary of commerce, David J. Rothkopf, explained in the May/June issue of *Foreign Policy*, the nation will need "regiments of geeks" — "pocket-protector brigades" who "will provide the software, systems, and analytical resources" to "close the gaps Mohammed Atta and his associates revealed."

Such ideas have provoked the ire of civil liberties groups, which fear that governments, corporations, and the police will misuse the new technology. Schneier's concerns are more basic. In his view, these measures can be useful, but their large-scale application will have little effect against terrorism. Worse, their use may make Americans less safe, because many of these tools fail badly — they're "brittle," in engineering jargon. Meanwhile, simple, effective, ductile measures are being overlooked or even rejected.

The distinction between ductile and brittle security dates back, Schneier has argued, to the nineteenth-century linguist and cryptographer Auguste Kerckhoffs, who set down what is now known as Kerckhoffs's principle. In good crypto systems, Kerckhoffs wrote, "the system should not depend on secrecy, and it should be able to fall into the enemy's hands without disadvantage." In other words, it should permit people to keep messages secret even if outsiders find out exactly how the encryption algorithm works.

At first blush this idea seems ludicrous. But contemporary cryptography follows Kerckhoffs's principle closely. The algorithms — the scrambling methods — are openly revealed; the only secret is the key. Indeed, Schneier says, Kerckhoffs's principle applies beyond codes and ciphers to security systems in general: every secret creates a potential failure point. Secrecy, in other words, is a prime cause of brittleness — and therefore something likely to make a system prone to catastrophic collapse. Conversely, openness provides ductility.

From this can be drawn several corollaries. One is that plans to add new layers of secrecy to security systems should automatically be viewed with suspicion. Another is that security systems that utterly depend on keeping secrets tend not to work very well. Alas, airport security is among these. Procedures for screening passengers, for examining luggage, for allowing people on the tarmac, for entering the cockpit, for running the autopilot software — all must be concealed, and all seriously compromise the system if they become known. As a result, Schneier wrote in the May issue of *Crypto-Gram*, brittleness "is an inherent property of airline security."

Few of the new airport security proposals address this problem. Instead, Schneier told me in Los Angeles, they address problems that don't exist. "The idea that to stop bombings cars have to park three hundred feet away from the terminal, but meanwhile they can drop off passengers right up front like they always have . . ." He laughed. "The only ideas I've heard that make any sense are reinforcing the cockpit door and getting the passengers to fight back." Both measures test well against Kerckhoffs's principle: knowing ahead of time that law-abiding passengers may forcefully resist a hijacking en masse, for example, doesn't help hijackers to fend off their assault. Both are small, compartmentalized measures that make the system more ductile, because no matter how hijackers get aboard, beefed-up doors and resistant passengers will make it harder for them to fly into a nuclear plant. And neither measure has any adverse effect on civil liberties.

Evaluations of a security proposal's merits, in Schneier's view, should not be much different from the ordinary cost-benefit calculations we make in daily life. The first question to ask of any new security proposal is: What problem does it solve? The second: What problems does it cause, especially when it fails?

Failure comes in many kinds, but two of the more important are simple failure (the security measure is ineffective) and what might be called subtractive failure (the security measure makes people less secure than before). An example of simple failure is face recognition technology. In basic terms, face recognition devices photograph people; break down their features into "facial building elements"; convert these into numbers that, like fingerprints, uniquely identify individuals; and compare the results with those stored in a database. If someone's facial score matches that of a criminal in the database, the person is detained. Since September 11, face recognition technology has been placed in an increasing number of public spaces: airports, beaches, nightlife districts. Even visitors to the Statue of Liberty now have their faces scanned.

Face recognition software could be useful. If an airline employee has to type in an identifying number to enter a secure area, for example, it can help to confirm that someone claiming to be that specific employee is indeed that person. But it cannot pick random terrorists out of the mob in an airline terminal. That much larger task requires comparing many sets of features with the many other sets of features in a database of people on a "watch list." Identix, of Minnesota, one of the largest face recognition technology companies, contends that in independent tests its FaceIt software has a success rate of 99.32 percent — that is, when the software matches a passenger's face with a face on a list of terrorists, it is mistaken only 0.68 percent of the time. Assume for the moment that this claim is credible; assume, too, that good pictures of suspected terrorists are readily available. About 25 million passengers used Boston's Logan Airport in 2001. Had face recognition software been used on 25 million faces, it would have wrongly picked out just 0.68 percent of them — but that would have been enough, given the large number of passengers, to flag as many as 170,000 innocent people as terrorists. With almost 500 false alarms a day, the face recognition system would quickly become something to ignore.

The potential for subtractive failure, different and more troublesome, is raised by recent calls to deploy biometric identification tools across the nation. Biometrics — "the only way to prevent identity fraud," according to the former senator Alan K. Simpson, of Wyoming — identifies people by precisely measuring their physical characteristics and matching them against a database. The photographs on driver's licenses are an early example, but engineers have developed many high-tech alternatives, some of them already mentioned: fingerprint readers, voiceprint recorders, retina or iris scanners, face recognition devices, hand-geometry assayers, even signature-geometry analyzers, which register pen pressure and writing speed as well as the appearance of a signature.

Appealingly, biometrics lets people be their own ID cards — no more passwords to forget! Unhappily, biometric measures are often implemented poorly. This past spring three reporters at *c't*, a German digital culture magazine, tested a face recognition system, an iris scanner, and nine fingerprint readers. All proved easy to outsmart. Even at the highest security setting, Cognitec's FaceVACS-Logon could be fooled by showing the sensor a short digital movie of someone known to the system — the president of a company, say — on a laptop screen. To beat Panasonic's Authenticam iris scanner, the German journalists photographed an authorized user, took the photo and created a detailed, life-size image of his eyes, cut out the pupils, and held the image up before their faces like a mask. The scanner read the iris, detected the presence of a human pupil — and accepted the imposture. Many of the fingerprint readers could be tricked simply by breathing on them, reactivating the last user's fingerprint. Beating the more sophisticated Identix Bio-Touch fingerprint reader required a trip to a hobby shop. The journalists used graphite powder to dust the latent fingerprint — the kind left on glass — of a previous, authorized user; picked up the image on adhesive tape; and pressed the tape on the reader. The Identix reader, too, was fooled. Not all biometric devices are so poorly put together, of course. But all of them fail badly.

Consider the legislation introduced in May by Congressmen Jim Moran and Tom Davis, both of Virginia, that would mandate biometric data chips in driver's licenses — a sweeping, nationwide data collection program, in essence. (Senator Dick Durbin, of Illinois, is proposing measures to force states to use a "single identifying designation unique to the individual on all driver's licenses"; President George W. Bush has already signed into law a requirement for biometric student visas.) Although Moran and Davis tied their proposal to the need for tighter security after last year's attacks, they also contended that the nation could combat fraud by using smart licenses with bank, credit, and Social Security cards, and for voter registration and airport identification. Maybe so, Schneier says. "But think about screw-ups, because the system *will* screw up."

Smart cards that store nonbiometric data have been routinely cracked in the past, often with inexpensive oscilloscope-like devices that detect and interpret the timing and power fluctuations as the chip operates. An even cheaper method, announced in May by two Cambridge security researchers, requires only a bright light, a standard microscope, and duct tape. Biometric ID cards are equally vulnerable. Indeed, as a recent National Research Council study points out, the extra security supposedly provided by biometric ID cards will raise the economic incentive to counterfeit or steal them, with potentially disastrous consequences to the victims. "Okay, somebody steals your thumbprint," Schneier says. "Because we've centralized all the functions, the thief can tap your credit, open your medical records, start your car, any number of things. Now what do you do? With a credit card, the bank can issue you a new card with a new number. But this is your *thumb* — you can't get a new one."

The consequences of identity fraud might be offset if biometric licenses and visas helped to prevent terrorism. Yet smart cards would not have stopped the terrorists who attacked the World Trade Center and the Pentagon. According to the FBI, all the hijackers seem to have been who they said they were; their intentions, not their identities, were the issue. Each entered the country with a valid visa, and each had a photo ID in his real name (some obtained their IDs fraudulently, but the fakes correctly identified them). "What problem is being solved here?" Schneier asks.

Good security is built in overlapping, cross-checking layers, to slow down attacks; it reacts limberly to the unexpected. Its most important components are almost always human. "Governments have been relying on intelligent, trained guards for centuries," Schneier says. "They spot people doing bad things and then use laws to arrest them. All in all, I have to say, it's not a bad system."

The Human Touch

One of the first times I met with Schneier was at the Cato Institute, a libertarian think tank in Washington, D.C., that had asked him to speak about security. Afterward I wondered how the Cato people had reacted to the speech. Libertarians love cryptography, because they believe that it will let people keep their secrets forever, no matter what a government wants. To them, Schneier was a kind of hero, someone who fought the good fight. As a cryptographer, he had tremendous street cred: he had developed some of the world's coolest ciphers, including the first rigorous encryption algorithm ever published in a best-selling novel (*Cryptonomicon*, by Neal Stephenson) and the encryption for the "virtual box tops" on Kellogg's cereals (children type a code from the box top into a Web site to win prizes), and had been one of the finalists in the competition to write algorithms for the federal government's new encryption standard, which it adopted last year. Now, in the nicest possible way, he had just told the libertarians the bad news: he still loved cryptography for the intellectual challenge, but it was not all that relevant to protecting the privacy and security of real people.

In security terms, he explained, cryptography is classed as a protective countermeasure. No such measure can foil every attack, and all attacks must still be both detected and responded to. This is particularly true for digital security, and Schneier spent most of his speech evoking the staggering insecurity of networked computers. Countless numbers are broken into every year, including machines in people's homes. Taking over computers is simple with the right tools, because software is so often misconfigured or flawed. In the first five months of this year, for example, Microsoft released five "critical" security patches for Internet Explorer, each intended to rectify lapses in the original code.

Computer crime statistics are notoriously sketchy, but the best of a bad lot come from an annual survey of corporations and other institutions by the FBI and the Computer Security Institute, a research and training organization in San Francisco. In the most recent survey, released in April, 90 percent of the respondents had detected one or more computer security breaches within the previous twelve months — a figure that Schneier calls "almost certainly an underestimate." His own experience suggests that a typical corporate network suffers a serious security breach four to six times a year — more often if the network is especially large or its operator is politically controversial.

Luckily for the victims, this digital mayhem is mostly wreaked not by the master hackers depicted in Hollywood techno-thrillers but by "script kiddies" — youths who know just enough about computers to download and run automated break-in programs. Twenty-four hours a day, seven days a week, script kiddies poke and prod at computer networks, searching for any of the thousands of known security vulnerabilities that administrators have not yet patched. A typical corporate network, Schneier says, is hit by such doorknob-rattling several times an hour. The great majority of these attacks achieve nothing, but eventually any existing security holes will be found and exploited. "It's very hard to communicate how bad the situation is," Schneier says, "because it doesn't correspond to our normal intuition of the world. To a first approximation, bank vaults are secure. Most of them don't get broken into because it takes real skill. Computers are the opposite. Most of them get broken into all the time, and it takes practically no skill." Indeed, as automated cracking software improves, it takes ever less knowledge to mount ever more sophisticated attacks.

Given the pervasive insecurity of networked computers, it is striking that nearly every proposal for "homeland security" entails the creation of large national databases. The Moran-Davis proposal, like other biometric schemes, envisions storing smart card information in one such database; the USA PATRIOT Act effectively creates another; the proposed Department of Homeland Security would "fuse and analyze" information from more than a hundred agencies and would "merge under one roof" scores or hundreds of previously separate databases. (A representative of the new department told me no one had a real idea of the number. "It's a lot," he said.) Better coordination of data could have obvious utility, as was made clear by recent headlines about the failure of the FBI and the CIA to communicate. But carefully linking selected fields of data is different from creating huge national repositories of information about the citizenry, as is being proposed. Larry Ellison, the CEO of Oracle, has dismissed cautions about such databases as whiny cavils that don't take into account the existence of murderous adversaries. But murderous adversaries are exactly why we should ensure that new security measures actually make American life safer.

Any new database must be protected, which automatically entails a new layer of secrecy. As Kerckhoffs's principle suggests, the new secrecy introduces a new failure point. Government information is now scattered through scores of databases; however inadvertently, it has been compartmentalized — a basic security practice. (Following this practice, tourists divide their money between their wallets and hidden pouches; pickpockets are less likely to steal it all.) Many new proposals would change that. An example is Attorney General John Ashcroft's plan, announced in June, to fingerprint and photograph foreign visitors "who fall into categories of elevated national security concern" when they enter the United States ("approximately 100,000" will be tracked this way in the first year). The fingerprints and photographs will be compared with those of "known or suspected terrorists" and "wanted criminals." Alas, no such database of terrorist fingerprints and photographs exists. Most terrorists are outside the country and thus hard to fingerprint, and latent fingerprints rarely survive bomb blasts. The databases of "wanted criminals" in Ashcroft's plan seem to be those maintained by the FBI and the Immigration and Naturalization

Service. But using them for this purpose would presumably involve merging computer networks in these two agencies with the visa procedure in the State Department — a security nightmare, because no one entity will fully control access to the system.

Equivalents of the big, centralized databases under discussion already exist in the private sector: corporate warehouses of customer information, especially credit card numbers. The record there is not reassuring. "Millions upon millions of credit card numbers have been stolen from computer networks," Schneier says. So many, in fact, that Schneier believes that everyone reading this article "has, in his or her wallet right now, a credit card with a number that has been stolen," even if no criminal has yet used it. Number thieves, many of whom operate out of the former Soviet Union, sell them in bulk: $1,000 for 5,000 credit card numbers, or twenty cents apiece. In a way, the sheer volume of theft is fortunate: so many numbers are floating around that the odds are small that any one will be heavily used by bad guys.

Large-scale federal databases would undergo similar assaults. The prospect is worrying, given the government's long-standing reputation for poor information security. Since September 11, at least forty government networks have been publicly cracked by typographically challenged vandals with names like "CriminalS," "S4t4n1c Souls," "cr1m3 org4n1z4do," and "Discordian Dodgers." Summing up the problem, a House subcommittee last November awarded federal agencies a collective computer security grade of F. According to representatives of Oracle, the federal government has been talking with the company about employing its software for the new central databases. But judging from the past, involving the private sector will not greatly improve security. In March, CERT/CC, a computer security watchdog based at Carnegie Mellon University, warned of thirty-eight vulnerabilities in Oracle's database software. Meanwhile, a centerpiece of the company's international advertising is the claim that its software is "unbreakable." Other software vendors fare no better: CERT/CC issues a constant stream of vulnerability warnings about every major software firm.

Schneier, like most security experts I spoke to, does not oppose consolidating and modernizing federal databases per se. To avoid creating vast new opportunities for adversaries, the overhaul should be incremental and on a small scale. Even so, it would need to be planned with extreme care — something that shows little sign of happening.

One key to the success of digital revamping will be a little-mentioned, even prosaic feature: training the users not to circumvent secure systems. The federal government already has several computer networks — INTELINK, SIPRNET, and NIPRNET among them — that are fully encrypted, accessible only from secure rooms and buildings, and never connected to the Internet. Yet despite their lack of Net access, the secure networks have been infected by e-mail perils such as the Melissa and I Love You viruses, probably because some official checked e-mail on a laptop, got infected, and then plugged the same laptop into the classified network. Because secure networks are unavoidably harder to work with, people are frequently tempted to bypass them — one reason that researchers at weapons labs sometimes transfer their files to insecure but more convenient machines.

Schneier has long argued that the best way to improve the very bad situation in computer security is to change software licenses. If software is blatantly unsafe, owners have no recourse, because it is licensed rather than bought, and the licenses forbid litigation. It is unclear whether the licenses can legally do this (courts currently disagree), but as a practical matter, it is next to impossible to win a lawsuit against a software firm. If some big software companies lose product liability suits, Schneier believes, their confreres will begin to take security seriously.

Computer networks are difficult to keep secure in part because they have so many functions, each of which must be accounted for. For that reason Schneier and other experts tend to favor narrowly focused security measures — more of them physical than digital — that target a few precisely identified problems. For air travel, along with reinforcing cockpit doors and teaching passengers to fight back, examples include armed, uniformed — *not* plainclothes — guards on select flights; "dead man" switches that in the event of a pilot's incapacitation force planes to land by autopilot at the nearest airport; positive bag matching (ensuring that luggage does not get on a plane unless its owner also boards); and separate decompression facilities that detonate any altitude bombs in cargo before takeoff. None of these is completely effective; bag matching, for instance, would not stop suicide bombers. But all are well tested, known to at least impede hijackers, not intrusive to passengers, and unlikely to make planes less secure if they fail.

It is impossible to guard all potential targets because anything and everything can be subject to attack. Palestinian suicide bombers have shown this by murdering at random the occupants of pool halls and hotel meeting rooms. Horrible as these incidents are, they do not risk the lives of thousands of people, as would attacks on critical parts of the national infrastructure: nuclear-power plants, hydroelectric dams, reservoirs, gas and chemical facilities. Here a classic defense is available: tall fences and armed guards. Yet this past spring the Bush administration cut by 93 percent the funds requested by the Energy Department to bolster security for nuclear weapons and waste; it denied completely the funds requested by the Army Corps of Engineers for guarding two hundred reservoirs, dams, and canals, leaving fourteen large public works projects with no budget for protection. A recommendation by the American Association of Port Authorities that the nation spend a total of $700 million to inspect and control ship cargo (today less than 2 percent of container traffic is inspected) has so far resulted in grants of just $92 million. In all three proposals, most of the money would have been spent on guards and fences.

The most important element of any security measure, Schneier argues, is people, not technology — and the people need to be at the scene. Recall the German journalists who fooled the fingerprint readers and iris scanners. None of their tricks would have worked if a reasonably attentive guard had been watching. Conversely, legitimate employees with bandaged fingers or scratched corneas will never make it through security unless a guard at the scene is authorized to overrule the machinery. Giving guards increased authority provides more opportunities for abuse, Schneier says, so the guards must be supervised carefully. But a system with more people who have more responsibility "is more robust," he observed in the June Crypto-Gram, "and the best way to make things work. (The U.S. Marine Corps understands this principle; it's the heart of their chain of command rules.)"

"The trick is to remember that technology can't save you," Schneier says. "We know this in our own lives. We realize that there's no magic antiburglary dust we can sprinkle on our cars to prevent them from being stolen. We know that car alarms don't offer much protection. The Club at best makes burglars steal the car next to you. For real safety, we park on nice streets where people notice if somebody smashes the window. Or we park in garages, where somebody watches the car. In both cases, people are the essential security element. You always build the system around people."

Looking for Trouble

After meeting Schneier at the Cato Institute, I drove with him to the Washington command post of Counterpane Internet Security. It was the first time in many months that he had visited either of his company's two operating centers (the other is in Silicon Valley). His absence had been due not to inattentiveness but to his determination to avoid the classic high-tech mistake of involving the alpha geek in day-to-day management. Besides, he lives in Minneapolis and the company headquarters are in Cupertino, California. (Why Minneapolis? I asked. "My wife lives there," he said. "It seemed polite.") With his partner, Tom Rowley, supervising day-to-day operations, Schneier constantly travels on Counterpane's behalf, explaining how the company manages computer security for hundreds of large and medium-sized companies. It does this mainly by installing human beings.

The command post was nondescript even by the bland architectural standards of exurban office complexes. Gaining access was like a pop quiz in security: How would the operations center recognize and admit its boss, who was there only once or twice a year? In this country, requests for identification are commonly answered with a driver's license. A few years ago Schneier devoted considerable effort to persuading the State of Illinois to issue him a driver's license that showed no picture, signature, or Social Security number. But Schneier's license serves as identification just as well as a license showing a picture and a signature — which is to say, not all that well. With or without a picture, with or without a biometric chip, licenses cannot be more than state-issued cards with people's names on them: good enough for social purposes, but never enough to assure identification when it is important. Authentication, Schneier says, involves something a person knows (a password or a PIN, say), has (a physical token, such as a driver's license or an ID bracelet), or is (biometric data). Security systems should use at least two of these; the Counterpane center employs all three. At the front door Schneier typed in a PIN and waved an iButton on his key chain at a sensor (iButtons, made by Dallas Semiconductor, are programmable chips embedded in stainless steel discs about the size and shape of a camera battery). We entered a waiting room, where Schneier completed the identification trinity by placing his palm on a hand-geometry reader.

Beyond the waiting room, after a purposely long corridor studded with cameras, was a conference room with many electrical outlets, some of which Schneier commandeered for his cell phone, laptop, BlackBerry, and battery packs. One side of the room was a dark glass wall. Schneier flicked a switch, shifting the light and theatrically revealing the scene behind the glass. It was a Luddite nightmare: an auditorium-like space full of desks, each with two computer monitors; all the desks faced a wall of high-resolution screens. One displayed streams of data from the "sentry" machines that Counterpane installs in its clients' networks. Another displayed images from the video cameras scattered around both this command post and the one in Silicon Valley.

On a visual level, the gadgetry overwhelmed the people sitting at the desks and watching over the data. Nonetheless, the people were the most important part of the operation. Networks record so much data about their usage that overwhelmed managers frequently turn off most of the logging programs and ignore the others. Among Counterpane's primary functions is to help companies make sense of the data they already have. "We turn the logs back on and monitor them," Schneier says. Counterpane researchers developed software to measure activity on client networks, but no software by itself can determine whether an unusual signal is a meaningless blip or an indication of trouble. That was the job of the people at the desks.

Highly trained and well paid, these people brought to the task a quality not yet found in any technology: human judgment, which is at the heart of most good security. Human beings do make mistakes, of course. But they can recover from failure in ways that machines and software cannot. The well-trained mind is ductile. It can understand surprises and overcome them. It fails well.

When I asked Schneier why Counterpane had such Darth Vaderish command centers, he laughed and said it helped to reassure potential clients that the company had mastered the technology. I asked if clients ever inquired how Counterpane trains the guards and analysts in the command centers. "Not often," he said, although that training is in fact the center of the whole system. Mixing long stretches of inactivity with short bursts of frenzy, the work rhythm of the Counterpane guards would have been familiar to police officers and firefighters everywhere. As I watched the guards, they were slurping soft drinks, listening to techno-death metal, and waiting for something to go wrong. They were in a protected space, looking out at a dangerous world. Sentries around Neolithic campfires did the same thing. Nothing better has been discovered since. Thinking otherwise, in Schneier's view, is a really terrible idea.

BILL MCKIBBEN

It's Easy Being Green

FROM *Mother Jones*

THE MORE I SURVEYED my new car, the happier I got. "New car" is one of those phrases that make Americans unreasonably happy to begin with. And this one — well, it was a particularly shiny metallic blue. Better yet, it was the first Honda Civic hybrid electric sold in the state of Vermont: I'd traded in my old Civic (40 miles to the gallon), and now the little screen behind the steering wheel was telling me that I was getting 50, 51, 52 miles to the gallon. Even better yet, I was doing nothing strange or difficult or conspicuously ecological. If you didn't know there was an electric motor assisting the small gas engine — well, you'd never know. The owner's manual devoted far more space to the air bags and the heating system. It didn't look goofily Jetsonish like Honda's first hybrid, the two-seater Insight introduced in 2000. Instead, it looked like a Civic, the most vanilla car ever produced. "Our goal was to make it look, for lack of a better word, normal," explained Kevin Bynoe, spokesman for American Honda.

And the happier I got, the angrier I got. Because, as the Honda and a raft of other recent developments powerfully proved, energy efficiency, energy conservation, and renewable energy are ready for prime time. No longer the niche province of incredibly noble backyard tinkerers distilling biodiesel from used vegetable oil or building homes from Earth rammed into tires, the equipment and attitudes necessary to radically transform our energy system are now mainstream enough for those of us too lazy or too busy to try anything that seems hard. And yet the switch toward sensible energy still isn't happening. A few weeks before I picked up my car, an overwhelming bipartisan vote in the Senate had rejected calls to increase the mileage of the nation's new car fleet by 2015 — to increase it to 36 mpg, not as good as the Civic I'd traded in to buy this hybrid. The administration was pressing ahead with its plan for more drilling and refining. The world was suffering the warmest winter in history as more carbon dioxide pushed global temperatures ever higher. And people were dying in conflicts across wide swaths of the world, the casualties — at least in some measure — of America's insatiable demand for energy.

In other words, the gap between what we could be doing and what we are doing has never been wider. Consider:

• The Honda I was driving was the third hybrid model easily available in this country, following in the tire tracks of the Insight and the Toyota Prius. They take regular gas, they require nothing in the way of special service, and they boast waiting lists. And yet Detroit, despite a decade of massive funding from the Clinton administration, can't sell you one. Instead, after September 11, the automakers launched a massive campaign (zero financing, red, white, and blue ads) to sell existing stock, particularly the gas-sucking SUVs that should by all rights come with their own little Saudi flags on the hood.

• Even greater boosts in efficiency can come when you build or renovate a home. Alex Wilson, editor of *Environmental Building News,* says the average American house may be 20 percent more energy efficient than it was two decades ago, but simple tweaks like better windows and bulkier insulation could save 30 to 50 percent more energy with "very little cost implication." And yet building codes do almost nothing to boost such technologies, and the Bush administration is fighting to roll back efficiency gains for some appliances that Clinton managed to push through. For instance, air-conditioner manufacturers recently won a battle in the Senate to let them get away with making their machines only 20 percent more efficient, not the 30 percent current law demands. The difference in real terms? Sixty new power plants across the country by 2030.

• Or consider electric generation. For a decade or two, environmentalists had their fingers crossed when they talked about renewables. It was hard to imagine most Americans really trading in their grid connection for backyard solar panels with their finicky batteries. But such trade-offs are less necessary by the day. Around the world, wind power is growing more quickly than any other form of energy — Denmark, Germany, Spain, and India all generate big amounts of their power from ultramodern wind turbines. But in this country, where the never-ending breeze across the High Plains could generate twice as much electricity as the country uses, progress has been extraordinarily slow. (North Dakota, the windiest state in the union, has exactly four turbines.) Wind power is finally beginning to get some serious attention from the energy industry, but the technology won't live up to its potential until politicians stop subsidizing fossil fuels and give serious boosts to the alternatives.

And not all those politicians are conservative, either. In Massachusetts, even some true progressives, like the gubernatorial candidate Robert Reich, can't bring themselves to endorse a big wind installation proposed for six miles off Cape Cod. They have lots of arguments, most of which boil down to NIVOMD (Not in View of My Deck), a position particularly incongruous since Cape Cod will sink quickly beneath the Atlantic unless every weapon in the fight against global warming is employed as rapidly as possible.

What really haunts energy experts is the sense that, for the first time since the oil shocks of the early 1970s, the nation could have rallied around the cause of energy conservation and renewable alternatives last fall. In the wake of September 11, they agree, the president could have announced a pair of national goals — capture Osama and free ourselves from the oil addiction that leaves us endlessly vulnerable. "President Bush's failure will haunt me for decades," says Alan Durning, president of Northwest Environment Watch. "Bush had a chance to advance, in a single blow, three pressing national priorities: national security, economic recovery, and environmental protection. All the stars were aligned." If only, says Brent Blackwelder, president of Friends of the Earth, Bush had set a goal, like JFK and the space program. "We could totally get off oil in three decades." Instead, the president used the crisis to push for drilling in the Arctic National Wildlife Refuge, a present to campaign contributors that would yield a statistically insignificant new supply ten years down the road.

It's not just new technologies that Bush could have pushed, of course. Americans were, at least for a little while, in the mood to do something, to make some sacrifice, to rally around some cause. In the words of Charles Komanoff, a New York energy analyst, "The choice is between love of oil and love of country," and at least "in the initial weeks after September 11, it seemed that Americans were awakening at last to the true cost of their addiction to oil." In an effort to take advantage of that political window, Komanoff published a booklet showing just how simple it would be to cut America's oil use by 5 or 10 percent — not over the years it will take for the new technologies to really kick in, but over the course of a few weeks and with only minor modifications to our way of life.

For instance, he calculated, we could save 7 percent of the gasoline we use simply by eliminating one car trip in fourteen. The little bit of planning required to make sure you visit the grocery store three times a week instead of four would leave us with endlessly more oil than sucking dry the Arctic. Indeed, Americans are so energy-profligate that even minor switches save significant sums — if half the drivers in two-car households switched just a tenth of their travel to their more efficient vehicle, we'd instantly save 1 percent of our oil. Keep the damn Explorer; just leave it in the driveway once a week and drive the Camry.

A similar menu of small changes — cutting back on one airplane trip in seven, turning down the thermostat two degrees, screwing in a few compact fluorescent bulbs — and all of a sudden our endlessly climbing energy usage begins to decline. Impossible? Americans won't do it? Look at California. With the threat of power shortages looming and with some clever incentives provided by government and utilities, Californians last year found an awful lot of small ways to save energy that really added up: 79 percent reported taking some steps, and a third of households managed to cut their electric use by more than 20 percent. Not by becoming a Third World nation (the state's economy continued to grow), not by living in caves, not by suffering — but by turning off the lights when they left the room. In just the first six months of 2001, the Colorado energy guru Amory Lovins pointed out recently, "customers wiped out California's previous five to ten years of demand growth." Now the same companies that were scrambling to build new plants for the Golden State a year ago are backing away from their proposals, spooked by the possibility of an energy glut.

It's only in Washington, in fact, that nobody gets it. If you go to Europe or Asia, you'll find nations increasingly involved in planning for a different energy future: Every industrial country but the United States signed on to the Kyoto agreement at the last international conference on global warming, and some of those nations may actually meet their targets for carbon dioxide reductions. The Dutch consumer demand for green power outstrips even the capacity of their growing wind farms, while the Germans have taken the logical step of raising taxes on carbon-based fuels and eliminating them on renewable sources. Reducing fossil fuel use is an accepted, inevitable part of the political process on the Continent, the same way that "fighting crime" is in this country, and Europeans look with growing disgust at the depth of our addiction — only the events of September 11 saved America from a wave of universal scorn when Bush backed away from the Kyoto pact.

And in state capitols and city halls around this country, local leaders are beginning to act as well. Voters in San Francisco last year overwhelmingly approved an initiative to require municipal purchases of solar and wind power; in Seattle, the mayor's office announced an ambitious plan to meet or beat the Kyoto targets within the confines of the city and four suburbs.

Perhaps such actions might be expected in San Francisco and Seattle. But in June of 2001, the Chicago city government signed a contract with Commonwealth Edison to buy 10 percent of its power from renewables, a figure due to increase to 20 percent in five years. And in Salt Lake City, of all places, Mayor Rocky Anderson announced on the opening day of the Winter Olympics that his city, too, was going to meet the Kyoto standards — already, in fact, crews were at work changing lightbulbs in street lamps and planning new mass transit.

Even many big American corporations have gone much further than the Bush administration. As Alex Wilson, the green building expert, points out, "Corporations are pretty good at looking at the bottom line, which is directly affected by operating costs. They're good with numbers." If you can make your product with half the energy, well, that's just as good as increasing sales — and if you can put a windmill on the cover of your annual report, that's gravy.

In short, what pretty much everyone outside the White House has realized is this: The great economic shift of this century will be away from fossil fuels and toward renewable energy. That shift will happen with or without George W. Bush — there are too many reasons, from environmental to economic to geopolitical necessity, for it not to. But American policy can slow down the transition, perhaps by decades, and that is precisely what the administration would like to see. They have two reasons: One is the enormous debt they owe to the backers of their political careers, those coal and oil and gas guys who dictated large sections of the new energy policy. Those industries want to wring every last penny from their mines, their drill rigs, and their refineries — and if those extra decades mean that the planet's temperature rises a few degrees, well, that's business.

The other reason is just as powerful, though — it's the fear that Americans will blame their leaders if prices for gas go up too quickly. It's not an idle fear — certainly it was shared by Bill Clinton, who did nothing to stem the nation's love affair with SUVs, and by Al Gore, who, during his presidential campaign, demanded that the Strategic Petroleum Reserve be opened to drive down prices at the pump. But that's what makes Bush's post-September silence on this issue so sad. For once a U.S. president had the chance to turn it all around — to say that this was a sacrifice we needed to make and one that any patriot would support. It's tragically likely he will have the same opportunity again in the years ahead, and tragically unlikely that he will take it.

In the meantime, there's work to be done in statehouses and city halls. And at the car lot — at least the ones with the Honda and Toyota signs out front. "This Civic has a slightly different front end and a roof-mounted antenna," says Honda's Bynoe. "But other than that, it looks like a regular Civic, and it drives like one too. It's not necessarily for hard-core enviros. You don't have to scream about it at the top of your lungs. It's just a car." But a very shiny blue. And I just came back from a trip to Boston: 59 miles to the gallon.

STEVE OLSON

The Royal We

FROM *The Atlantic Monthly*

A FEW YEARS AGO the Genealogical Office in Dublin moved from a back room of the Heraldic Museum up the street to the National Library. The old office wasn't big enough for all the people stopping by to track down their Irish ancestors, and even the new, much larger office is often crowded. Because of its history of oppression and Catholic fecundity, Ireland has been a remarkably productive exporter of people. The population of the island has never exceeded 10 million, but more than 70 million people worldwide claim Irish ancestry. On warm summer days, as tourists throng nearby Trinity College and Dublin Castle, the line of visitors waiting to consult one of the office's professional genealogists can stretch out the door.

I suspect that many people have had a fling with genealogy somewhat like mine. In my office I have a file containing the scattered lines of Olsons and Taylors, Richmans and Sigginses (my Irish ancestors), which I gathered several years ago in a paroxysm of family-mindedness. For the most part my ancestors were a steady stream of farmers, ministers, and malcontents. Yet a few of the Old World lines hint at something grander — they include a couple of knights and even a baron. I've never taken the trouble to find out, but I bet with a little work I could achieve that nirvana of genealogical research, demonstrated descent from a royal family.

Earlier this year I went to Dublin to learn more about the Irish side of my family and to talk about genealogy with Mark Humphrys, a young computer scientist at Dublin City University. Humphrys has dark hair, deep blue eyes, heavily freckled arms, and a pasty complexion. He became interested in genealogy as a teen-ager, after hearing romantic stories about his ancestors' roles in re-bellions against the English. But when he tried to trace his family further into the past, the trail ran cold. The Penal Laws imposed by England in the early eighteenth century forbade Irish Catholics from buying land or joining professions, which meant that very few permanent records of their existence were generated. "Irish peo-ple of Catholic descent are almost completely cut off from the past," Humphrys told me, as we sat in his office overlooking a busy construction site. (Dublin City University, which specializes in in-formation technology and the life sciences, is growing as rapidly as the northern Dublin suburb in which it is located.) "The great irony about Ireland is that even though we have this long, rich his-tory, almost no person of Irish-Catholic descent can directly con-nect to that history."

While a graduate student at Cambridge University, Humphrys fell in love with and married an Englishwoman, and investigating her genealogy proved more fruitful. Her family knew that they were descended from an illegitimate son of the tenth Earl of Pem-broke. After just a couple of hours in the Cambridge library, Hum-phrys showed that the Earl of Pembroke was a direct descendant of Edward III, making Humphrys's wife the king's great-granddaugh-ter twenty generations removed. Humphrys began to gather other genealogical tidbits related to English royalty. Many of the famous Irish rebels he'd learned about in school turned out to have ances-tors who had married into prominent Protestant families, which meant they were descended from English royalty. The majority of American presidents were also of royal descent, as were many of the well-known families of Europe.

Humphrys began to notice something odd. Whenever a reli-able family tree was available, almost anyone of European ancestry turned out to be descended from English royalty — even such un-likely people as Hermann Göring and Daniel Boone. Humphrys began to think that such descent was the rule rather than the ex-ception in the Western world, even if relatively few people had the documents to demonstrate it.

Humphrys began compiling his family genealogies in the 1980s, first on paper and then using computers. But he did much of his work on royal genealogies in the mid-1990s, when the World Wide

Web was just coming into general use. He began to put his findings on Web pages, with hyperlinks connecting various lines of descent. Suddenly dense networks of ancestry jumped out at him.
"I'd known these descents were interconnected, but I'd never
known how much," he told me. "You can't see the connections
reading the printed genealogies, because it's so hard to jump from
tree to tree. The problem is that genealogies aren't two-dimensional, so any attempt to put them on paper is more or less doomed
from the start. They aren't three-dimensional, either, or you could
make a structure. They have hundreds of dimensions."

Much of Humphrys's genealogical research now appears on his
Web page Royal Descents of Famous People. Sitting in his office, I
asked him to show me how it works. He clicked on the name Walt
Disney. Up popped a genealogy done by Brigitte Gastel Lloyd
(Humphrys links to the work of others whenever possible) showing
the twenty-two generations separating Disney from Edward I. Humphrys pointed at the screen. "Here we have a sir, so this woman is
the daughter of a knight. Maybe this woman will marry nobility, but
there's a limited pool of nobility, so eventually someone here is going to marry someone who's just wealthy. Then one of their children could marry someone who doesn't have that much money. In
ten generations you can easily get from princess to peasant."

The idea that virtually anyone with a European ancestor descends
from English royalty seems bizarre, but it accords perfectly with
some recent research done by Joseph Chang, a statistician at Yale
University. The mathematics of our ancestry is exceedingly complex, because the number of our ancestors increases exponentially,
not linearly. These numbers are manageable in the first few generations — two parents, four grandparents, eight great-grandparents,
sixteen great-great-grandparents — but they quickly spiral out of
control. Go back forty generations, or about a thousand years, and
each of us theoretically has more than a trillion direct ancestors —
a figure that far exceeds the total number of human beings who
have ever lived.

In a 1999 paper titled "Recent Common Ancestors of All Pres-
ent-Day Individuals," Chang showed how to reconcile the potentially huge number of our ancestors with the quantities of people
who actually lived in the past. His model is a mathematical proof

that relies on such abstractions as Poisson distributions and Markov chains, but it can readily be applied to the real world. Under the conditions laid out in his paper, the most recent common ancestor of every European today (except for recent immigrants to the Continent) was someone who lived in Europe in the surprisingly recent past — only about six hundred years ago. In other words, all Europeans alive today have among their ancestors the same man or woman who lived around 1400. Before that date, according to Chang's model, the number of ancestors common to all Europeans today increased until, about a thousand years ago, a peculiar situation prevailed: 20 percent of the adult Europeans alive in 1000 would turn out to be the ancestors of no one living today (that is, they had no children or all their descendants eventually died childless); each of the remaining 80 percent would turn out to be a direct ancestor of every European living today.

Chang's model incorporates one crucial assumption: random mating in the part of the world under consideration. For example, every person in Europe would have to have an equal chance of marrying every other European of the opposite sex. As Chang acknowledges in his paper, random mating clearly does not occur in reality; an Englishman is much likelier to marry a woman from England than a woman from Italy, and a princess is much likelier to marry a prince than a pauper. These departures from randomness must push back somewhat the date of Europeans' most recent common ancestor.

But Humphrys's Web page suggests that over many generations, mating patterns may be much more random than expected. Social mobility accounts for part of the mixing — what Voltaire called the slippered feet going down the stairs as the hobnailed boots ascend them. At the same time, revolutions overturn established orders, countries invade and colonize other countries, and people sometimes choose mates from far away rather than from next door. Even the world's most isolated peoples — Pacific islanders, for example — continually exchange potential mates with neighboring groups.

This constant churning of people makes it possible to apply Chang's analysis to the world as a whole. For example, almost everyone in the New World must be descended from English royalty — even people of predominantly African or Native American ancestry, because of the long history of intermarriage in the Americas. Similarly, everyone of European ancestry must descend from Muhammad. The line of descent for which records exist is through the daughter of the emir of Seville, who is reported to have converted from Islam to Catholicism in about 1200. But many other, unrecorded descents must also exist.

Chang's model has even more dramatic implications. Because people are always migrating from continent to continent, networks of descent quickly interconnect. This means that the most recent common ancestor of all 6 billion people on Earth today probably lived just a couple of thousand years ago. And not long before that, the majority of the people on the planet were the direct ancestors of everyone alive today. Confucius, Nefertiti, and just about any other ancient historical figure who was even moderately prolific must today be counted among everyone's ancestors.

Toward the end of our conversation, Humphrys pointed out something I hadn't considered. The same process works going forward in time; in essence, every one of us who has children and whose line does not go extinct is suspended at the center of an immense genetic hourglass. Just as we are descended from most of the people alive on the planet a few thousand years ago, several thousand years hence, each of us will be an ancestor of the entire human race — or of no one at all.

The dense interconnectedness of the human family might seem to take some of the thrill out of genealogical research. Sure, I was able to show in the Genealogical Office that my Siggins ancestors are descended from the fourteenth-century Syggens of County Wexford; but I'm also descended from most of the other people who lived in Ireland in the fourteenth century. Humphrys took issue with my disillusionment. It's true that everyone's roots go back to the same family tree, he said. But each path to our common past is different, and reconstructing that path, using whatever records are available, is its own reward. "You can ask whether everyone in the Western world is descended from Charlemagne, and the answer is yes, we're all descended from Charlemagne. But can you prove it? That's the game of genealogy."

DENNIS OVERBYE

A New View of Our Universe

Only One of Many

FROM *The New York Times*

ASTRONOMERS HAVE GAZED OUT at the universe for centuries, asking why it is the way it is. But lately a growing number of them are dreaming of universes that never were and asking, why not?

Why, they ask, do we live in three dimensions of space and not two, ten, or twenty-five? Why is a light ray so fast and a whisper so slow? Why are atoms so tiny and stars so big? Why is the universe so old? Does it have to be that way, or are there places, other universes, where things are different?

Once upon a time (only a century ago), a few billion stars and gas clouds smeared along the Milky Way were thought to encompass all of existence, and the notion of understanding it was daunting — and hubristic — enough. Now astronomers know that galaxies are scattered like dust across the cosmos. And understanding them might require recourse to an even broader canvas, what they sometimes call a "multiverse."

For some cosmologists, that means universes sprouting from one another in an endless geometric progression, like mushrooms upon mushrooms upon mushrooms, or baby universes hatched inside black holes. Others imagine island universes floating and even colliding in a fifth dimension.

For example, Dr. Max Tegmark, a University of Pennsylvania cosmologist, has posited at least four different levels of universes, ranging from the familiar (impossibly distant zones of our own universe) to the strange (space-times in which the fundamental laws of physics are different).

Dr. Martin Rees, a University of Cambridge cosmologist and the Astronomer Royal, said contemplating these alternate universes could help scientists distinguish which features of our own universe are fundamental and necessary and which are accidents of cosmic history. "It's all science, but science for the twenty-first century, to seek the answers to these questions," Dr. Rees said, adding that he is often accused of believing in other universes. "I don't believe," he said, "but I think it's part of science to find out."

Some cosmologists now say the realm we call the observable universe — roughly 14 billion light-years deep of galaxies and stars — could be only a small patch of a vast bubble or "pocket" in a much vaster ensemble bred endlessly in a chain of big bangs.

The idea, they say, is a natural extension of the theory of inflation, introduced by Dr. Alan Guth, now at the Massachusetts Institute of Technology, in 1980. That theory asserts that when the universe was less than a trillionth of a trillionth of a second old, it underwent a brief hyperexplosive growth spurt fueled by an antigravitational force embedded in space itself, a possibility suggested by theories of modern particle physics.

Because inflation can grow a whole universe from about an ounce of primordial stuff, Dr. Guth likes to refer to the universe as "the ultimate free lunch." But Dr. Guth and various other theorists — including Dr. Andrei Linde of Stanford, Dr. Alexander Vilenkin of Tufts, and Dr. Paul Steinhardt of Princeton — have suggested that it may be an endless one as well. Once inflation starts anywhere, it will keep happening over and over again, they say, spawning a chain of universes, bubbles within bubbles, in a scheme that Dr. Linde called "eternal inflation."

"Once you've discovered it's easy to make a universe out of an ounce of vacuum, why not make a bunch of them?" asked Dr. Craig Hogan, a cosmologist at the University of Washington.

In fact, Dr. Guth said, "Inflation pretty much forces the idea of multiple universes upon us."

Moreover, there is no reason to expect that these universes will be identical. Even within our own bubble, tiny random nonuniformities in the primordial raw material would cause the cosmos to look different from place to place. If the universe is big enough, Dr. Tegmark and others say, everything that can happen will happen, so that if we could look out far enough we would eventually discover an exact replica of ourselves.

Moreover, cosmologists say, the laws of physics themselves, as experienced by creatures like ourselves, confined to four dimensions and the energy scales of ordinary life, could evolve differently in different bubble universes.

"Geography is a now a much more interesting subject than you thought," Dr. John Barrow, a physicist at the University of Cambridge, observed.

Inflation has gained much credit with cosmologists despite its strangeness, Dr. Guth noted, because it plays a vital role in calculations of the Big Bang that have been vindicated by the detection of the radio waves it produced. The prediction of other universes must therefore be taken seriously, he said.

The prospect of this plethora of universes has brought new attention to a philosophical debate that has lurked on the edges of science for the last few decades, a debate over the role of life in the universe and whether its physical laws are unique — or, as Einstein once put it, "whether God had any choice."

Sprinkled through the Standard Model, the suite of equations that describe all natural phenomena, are various mysterious constants, like the speed of light or the masses of the elementary particles, whose value is not specified by any theory now known.

In effect, the knobs on nature's console have been set to these numbers. Scientists can imagine twiddling them, but it turns out that nature is surprisingly finicky, they say, and only a narrow range of settings is suitable for the evolution of complexity or Life as We Know It.

For example, much of the carbon and oxygen needed for life is produced by the fusion of helium atoms in stars called red giants. But a change of only half a percent in the strength of the so-called strong force that governs nuclear structure would be enough to prevent those reactions from occurring, according to recent work by Dr. Heinz Oberhummer of Vienna University of Technology. The result would be a dearth of the raw materials of biology, he said.

Similarly, a number known as the fine structure constant characterizes the strength of electromagnetic forces. If it were a little larger, astronomers say, stars could not burn, and if it were only a little smaller, molecules would never form.

So is this a lucky universe, or what?

In 1974 Dr. Brandon Carter, a theoretical physicist then at Cambridge, now at the Paris Observatory in Meudon, pointed out that these coincidences were not just luck but were rather necessary preconditions for us to be looking at the universe. After all, we are hardly likely to discover laws that are incompatible with our own existence.

That insight is the basis of what Dr. Carter called the anthropic principle, an idea that means many things to many scientists. Expressed most emphatically, it declares that the universe is somehow designed for life. Or as the physicist Freeman Dyson once put it, "The universe in some sense must have known that we were coming."

This notion horrifies some physicists, who feel it is their mission to find a mathematical explanation of nature that leaves nothing to chance or "the whim of the Creator," in Einstein's phrase. "It touches on philosophical issues that scientists oftentimes skirt," said Dr. John Schwarz, a physicist and string theorist at the California Institute of Technology. "There should be mathematical ways of understanding how nature works."

Dr. Steven Weinberg, the University of Texas physicist and Nobel laureate, referred to this so-called strong version of the anthropic principle as "little more than mystical mumbo jumbo" in a recent article in the *New York Review of Books*.

Nevertheless, the "A word" is popping up more and more lately, at conferences and in the scientific literature, often to the groans of particle physicists. The reason is the multiverse.

"It is possible that as theoretical physics develops, it will present us with multiple universes," Dr. Weinberg said.

If different laws or physical constants prevail in other bubble universes, the conditions may not allow the existence of life or intelligence, he explained. In that case the anthropic principle loses its mysticism and simply becomes a prescription for deciding which bubbles are capable of supporting life.

But many string theorists still resent the principle as an abridgment on their ambitions. The result has been a spirited debate about what physicists can expect from their theories.

"They have the pious hope that string theory will uniquely determine all the constants of nature," said Dr. Barrow, who wrote the

1984 book *The Cosmological Anthropic Principle* with Dr. Frank Tipler, a Tulane University physicist. The book argued that once life emerges in the universe it will never die.

In a recent paper titled "The Beginning of the End of the Anthropic Principle," three physicists — Dr. Gordon Kane of the University of Michigan and Dr. Malcolm Perry and Dr. Anna Zytkow, both of Cambridge — argued that a unified theory of physics, as string theory purports to be, when finally formulated, would specify most of the constants of nature or specify relationships between them, leaving little room for anthropic arguments. "The anthropic principle isn't as anthropic as people wanted," Dr. Kane said in an interview.

But in a rejoinder titled "Why the Universe Is Just So," Dr. Hogan of Washington argued that physics was replete with messy processes like quantum effects, which leave some aspects of reality and the laws of physics to chance. According to string theory, he pointed out, the laws of physics that we mortals experience are low-energy, four-dimensional shadows, of sorts, of a ten- or eleven-dimensional universe. As a result, the so-called fundamental constants could look different in different bubbles.

Dr. Hogan admitted that this undermined some of the traditional aspirations of physics, writing, "At least some properties of the world might not have an elegant mathematical explanation, and we can try to guess which ones these are."

Even string theorists like Dr. Kane admit that, in the absence of a final form of the theory, they have no idea how many solutions there may be — one, many, or even an infinite number — to the "final" string equations. Each one would correspond to a different condition of space-time, with a different set of physical constants. "Any set that allows life to happen will have life," he said.

But even some of the most hard-core physicists, including Dr. Weinberg, suggest they may have to resort to the anthropic principle to explain one of the deepest mysteries looming like a headache over science: the discovery that the expansion of the universe seems to be speeding up, perhaps in a kind of low-energy reprise of an inflation episode 14 billion years ago.

Cosmologists suspect that a repulsion or antigravity associated with space itself is propelling this motion. This force, known as the cosmological constant, was first proposed by Einstein back in 1917 and has been a problem ever since — "a veritable crisis," Dr. Weinberg has called it.

According to astronomical observations, otherwise undetectable energy — "dark energy" — accounts for about two-thirds of the mass-energy of the universe today, outweighing matter by two to one. But according to modern quantum physics, empty space should be seething with energy that would outweigh matter in the universe by far, far more, by a factor of at least 10^{60}. This mismatch has been called the worst discrepancy in the history of physics.

But that mismatch is crucial for life, as Dr. Weinberg first pointed out in 1987. At the time there was no evidence for a cosmological constant, and many physicists presumed that its magnitude was in fact zero.

In his paper, Dr. Weinberg used so-called anthropic reasoning to pin the value of any cosmological constant to between about one-tenth and a few times the density of matter in the universe. If it were any larger, he said, the universe would blow apart before galaxies had a chance to form, leaving no cradle for the stellar evolution of elements necessary for life or other complicated structures.

The measured value of the constant is about what would be expected from anthropic arguments, Dr. Weinberg said, adding that nobody knows enough about physics yet to tell whether there are other universes with other constants. He called the anthropic principle "a sensible approach" to the cosmological constant problem.

"We may wind up using the anthropic principle to satisfy our sense of wonder about why things are the way they are," Dr. Weinberg said.

For Dr. Rees, the Astronomer Royal, it is not necessary to observe other universes to gain some confidence that they may exist. One thing that will help, he explained, is a more precise theory of how the cosmological constant may vary and how it will affect life in the universe. We should live in a statistically typical example of the range of universes compatible with life, he explained. For example, if the cosmological constant was, say, 10 percent of the maximum value consistent with life, that would be acceptable, he said. "If it was a millionth, that would raise eyebrows."

Another confidence builder would be more support for the theory of inflation, either in the form of evidence from particle physics theory or measurements of the cosmic Big Bang radiation that gave a more detailed model of what theoretically happened during that first trillionth of a trillionth of a second. "If we had a theory, then we would know whether there were many big bangs or one, and then we would know if the features we see are fixed laws of the universe or bylaws for which we can never have an explanation," Dr. Rees said.

In a talk last month at a cosmology conference in Chicago, Dr. Joseph Polchinski, from the Institute for Theoretical Physics at the University of California in Santa Barbara, speculated that there could be 10^{60} different solutions to the basic string equations, thus making it more likely that at least one universe would have a friendly cosmological constant.

Reminded that he had once joked about retiring if a cosmological constant was discovered, on the ground that the dreaded anthropic principle would be the only explanation, he was at first at a loss for words. Later he said he hoped the range of solutions and possible universes permitted by string theory could be narrowed by astronomical observations and new theoretical techniques to the point where the anthropic principle could be counted out as an explanation. "Life is still good," he said.

But Dr. Hogan said that multiple universes would have to be taken seriously if they came out of equations that science had faith in. "You have to be open-minded," he said. "You can't impose conditions.

"It's the most scientific attitude," he added.

STEVEN PINKER

The Blank Slate

FROM *Discover*

IF YOU READ the pundits in newspapers and magazines, you may have come across some remarkable claims about the malleability of the human psyche. Here are a few from my collection of clippings:

- Little boys quarrel and fight because they are encouraged to do so.
- Children enjoy sweets because their parents use them as rewards for eating vegetables.
- Teenagers get the idea to compete in looks and fashion from spelling bees and academic prizes.
- Men think the goal of sex is an orgasm because of the way they were socialized.

If you find these assertions dubious, your skepticism is certainly justified. In all cultures little boys quarrel, children like sweets, teens compete for status, and men pursue orgasms without the slightest need of encouragement or socialization. In each case, the writers made their preposterous claims without a shred of evidence — without even a nod to the possibility that they were saying something common sense might call into question.

Intellectual life today is beset with a great divide. On one side is a militant denial of human nature, a conviction that the mind of a child is a blank slate that is subsequently inscribed by parents and society. For much of the past century, psychology has tried to explain all thought, feeling, and behavior with a few simple mechanisms of learning by association. Social scientists have tried to explain all customs and social arrangements as a product of the surrounding culture. A long list of concepts that would seem natural to the human way of thinking — emotions, kinship, the sexes — are said to have been "invented" or "socially constructed."

At the same time, there is a growing realization that human nature won't go away. Anyone who has had more than one child, or been in a heterosexual relationship, or noticed that children learn language but house pets don't has recognized that people are born with certain talents and temperaments. An acknowledgment that we humans are a species with a timeless and universal psychology pervades the writings of great political thinkers, and without it we cannot explain the recurring themes of literature, religion, and myth. Moreover, the modern sciences of mind, brain, genes, and evolution are showing that there is something to the commonsense idea of human nature. Although no scientist denies that learning and culture are crucial to every aspect of human life, these processes don't happen by magic. There must be complex innate mental faculties that enable human beings to create and learn culture.

Sometimes the contradictory attitudes toward human nature divide people into competing camps. The blank slate camp tends to have greater appeal among those in the social sciences and humanities than it does among biological scientists. And until recently, it was more popular on the political left than it was on the right.

But sometimes both attitudes coexist uneasily inside the mind of a single person. Many academics, for example, publicly deny the existence of intelligence. But privately, academics are *obsessed* with intelligence, discussing it endlessly in admissions, in hiring, and especially in their gossip about one another. And despite their protestations that it is a reactionary concept, they quickly invoke it to oppose executing a murderer with an I.Q. of 64 or to support laws requiring the removal of lead paint because it may lower a child's I.Q. by five points. Similarly, those who argue that gender differences are a reversible social construction do not treat them that way in their advice to their daughters, in their dealings with the opposite sex, or in their unguarded gossip, humor, and reflections on their lives.

No good can come from this hypocrisy. The dogma that human nature does not exist, in the face of growing evidence from science and common sense that it does, has led to contempt among many scholars in the humanities for the concepts of evidence and truth. Worse, the doctrine of the blank slate often distorts science itself by making an extreme position — that culture alone determines behavior — seem moderate, and by making the moderate position — that behavior comes from an interaction of biology and culture — seem extreme.

For example, many policies on parenting come from research that finds a correlation between the behavior of parents and of their children. Loving parents have confident children, authoritative parents (neither too permissive nor too punitive) have well-behaved children, parents who talk to their children have children with better language skills, and so on. Thus everyone concludes that parents should be loving, authoritative, and talkative, and if children don't turn out well, it must be the parents' fault.

Those conclusions depend on the belief that children are blank slates. It ignores the fact that parents provide their children with genes, not just an environment. The correlations may be telling us only that the same genes that make adults loving, authoritative, and talkative make their children self-confident, well behaved, and articulate. Until the studies are redone with adopted children (who get only their environment from their parents), the data are compatible with the possibility that genes make all the difference, that parenting makes all the difference, or anything in between. Yet the extreme position — that parents are everything — is the only one researchers entertain.

The denial of human nature has not just corrupted the world of intellectuals but has harmed ordinary people. The theory that parents can mold their children like clay has inflicted child-rearing regimes on parents that are unnatural and sometimes cruel. It has distorted the choices faced by mothers as they try to balance their lives, and it has multiplied the anguish of parents whose children haven't turned out as hoped. The belief that human tastes are reversible cultural preferences has led social planners to write off people's enjoyment of ornament, natural light, and human scale and forced millions of people to live in drab cement boxes. And the conviction that humanity could be reshaped by massive social engineering projects has led to some of the greatest atrocities in history.

The phrase "blank slate" is a loose translation of the medieval Latin term *tabula rasa* — scraped tablet. It is often attributed to the seventeenth-century English philosopher John Locke, who wrote that the mind is "white paper void of all characters." But it became the official doctrine among thinking people only in the first half of the twentieth century, as part of a reaction to the widespread belief in the intellectual or moral inferiority of women, Jews, nonwhite races, and non-Western cultures.

Part of the reaction was a moral repulsion from discrimination, lynchings, forced sterilizations, segregation, and the Holocaust. And part of it came from empirical observations. Waves of immigrants from southern and eastern Europe filled the cities of America and climbed the social ladder. African Americans took advantage of "Negro colleges" and migrated northward, beginning the Harlem Renaissance. The graduates of women's colleges launched the first wave of feminism. To say that women and minority groups were inferior contradicted what people could see with their own eyes.

Academics were swept along by the changing attitudes, but they also helped direct the tide. The prevailing theories of mind were refashioned to make racism and sexism as untenable as possible. The blank slate became sacred scripture. According to the doctrine, any differences we see among races, ethnic groups, sexes, and individuals come not from differences in their innate constitution but from differences in their experiences. Change the experiences — by reforming parenting, education, the media, and social rewards — and you can change the person. Also, if there is no such thing as human nature, society will not be saddled with such nasty traits as aggression, selfishness, and prejudice. In a reformed environment, people can be prevented from learning these habits.

In psychology, behaviorists like John B. Watson and B. F. Skinner simply banned notions of talent and temperament, together with all the other contents of the mind, such as beliefs, desires, and feelings. This set the stage for Watson's famous boast: "Give me a dozen healthy infants, well-formed, and my own specified world to bring them up in, and I'll guarantee to take any one at random and train him to become any type of specialist I might select — doctor, lawyer, artist, merchant-chief, and yes, even beggar-man and thief, regardless of his talents, penchants, tendencies, abilities, vocations, and race of his ancestors."

Watson also wrote an influential child-rearing manual recommending that parents give their children minimum attention and love. If you comfort a crying baby, he wrote, you will reward the baby for crying and thereby increase the frequency of crying behavior.

In anthropology, Franz Boas wrote that differences among human races and ethnic groups come not from their physical constitution but from their *culture*. Though Boas himself did not claim that people were blank slates — he only argued that all ethnic groups are endowed with the same mental abilities — his students, who came to dominate American social science, went further. They insisted not just that *differences* among ethnic groups must be explained in terms of culture (which is reasonable) but that *every aspect* of human existence must be explained in terms of culture (which is not). "Heredity cannot be allowed to have acted any part in history," wrote Alfred Kroeber. "With the exception of the instinctoid reactions in infants to sudden withdrawals of support and to sudden loud noises, the human being is entirely instinctless," wrote Ashley Montagu.

In the second half of the twentieth century, the ideals of the social scientists of the first half enjoyed a well-deserved victory. Eugenics, social Darwinism, overt expressions of racism and sexism, and official discrimination against women and minorities were on the wane, or had been eliminated, from the political and intellectual mainstream in Western democracies.

At the same time, the doctrine of the blank slate, which had been blurred with ideals of equality and progress, began to show cracks. As new disciplines such as cognitive science, neuroscience, evolutionary psychology, and behavioral genetics flourished, it became clearer that thinking is a biological process, that the brain is not exempt from the laws of evolution, that the sexes differ above the neck as well as below it, and that people are not psychological clones. Here are some examples of the discoveries.

Natural selection tends to homogenize a species into a standard design by concentrating the effective genes and winnowing out the ineffective ones. This suggests that the human mind evolved with a universal complex design. Beginning in the 1950s, the linguist Noam Chomsky of the Massachusetts Institute of Technology argued that a language should be analyzed not in terms of the list of sentences people utter but in terms of the mental computations that enable them to handle an unlimited number of new sentences in the language. These computations have been found to conform to a universal grammar. And if this universal grammar is embodied in the circuitry that guides babies when they listen to speech, it could explain how children learn language so easily.

Similarly, some anthropologists have returned to an ethnographic record that used to trumpet differences among cultures and have found an astonishingly detailed set of aptitudes and tastes that all cultures have in common. This shared way of thinking, feeling, and living makes all of humanity look like a single tribe, which the anthropologist Donald Brown of the University of California at Santa Barbara has called the universal people. Hundreds of traits, from romantic love to humorous insults, from poetry to food taboos, from exchange of goods to mourning the dead, can be found in every society ever documented.

One example of a stubborn universal is the tangle of emotions surrounding the act of love. In all societies, sex is at least somewhat "dirty." It is conducted in private, pondered obsessively, regulated by custom and taboo, the subject of gossip and teasing, and a trigger for jealous rage. Yet sex is the most concentrated source of physical pleasure granted by the nervous system. Why is it so fraught with conflict? For a brief period in the 1960s and 1970s, people dreamed of an erotopia in which men and women could engage in sex without hang-ups and inhibitions. "If you can't be with the one you love, love the one you're with," sang Stephen Stills. "If you love somebody, set them free," sang Sting.

But Sting also sang, "Every move you make, I'll be watching you." Even in a time when, seemingly, anything goes, most people do not partake in sex as casually as they partake in food or conversation. The reasons are as deep as anything in biology. One of the hazards of sex is a baby, and a baby is not just any seven-pound object but, from an evolutionary point of view, our reason for being. Every time a woman has sex with a man, she is taking a chance at sentencing herself to years of motherhood, and she is forgoing the opportunity to use her finite reproductive output with some other man. The man, for his part, may be either implicitly committing his sweat and toil to the incipient child or deceiving his partner about such intentions.

On rational grounds, the volatility of sex is a puzzle, because in an era with reliable contraception, these archaic entanglements should have no claim on our feelings. We should be loving the one we're with, and sex should inspire no more gossip, music, fiction, raunchy humor, or strong emotions than eating or talking does. The fact that people are tormented by the Darwinian economics of babies they are no longer having is testimony to the long reach of human nature.

Although the minds of normal human beings work in pretty much the same way, they are not, of course, identical. Natural selection reduces genetic variability but never eliminates it. As a result, nearly every one of us is genetically unique. And these differences in genes make a difference in mind and behavior, at least quantitatively. The most dramatic demonstrations come from studies of the rare people who *are* genetically identical, identical twins.

Identical twins think and feel in such similar ways that they sometimes suspect they are linked by telepathy. They are similar in verbal and mathematical intelligence, in their degree of life satisfaction, and in personality traits such as introversion, agreeableness, neuroticism, conscientiousness, and openness to experience. They have similar attitudes toward controversial issues such as the death penalty, religion, and modern music. They resemble each other not just in paper-and-pencil tests but in consequential behavior such as gambling, divorcing, committing crimes, getting into accidents, and watching television. And they boast dozens of shared idiosyncrasies such as giggling incessantly, giving interminable answers to simple questions, dipping buttered toast in coffee, and, in the case of Abigail van Buren and the late Ann Landers, writing indistinguishable syndicated advice columns. The crags and valleys of their electroencephalograms (brain waves) are as alike as those of a single person recorded on two occasions, and the wrinkles of their brains and the distribution of gray matter across cortical areas are similar as well.

Identical twins (who share all their genes) are far more similar than fraternal twins (who share just half their genes). This is as true when the twins are separated at birth and raised apart as when they are raised in the same home by the same parents. Moreover, biological siblings, who also share half their genes, are far more similar than adoptive siblings, who share no more genes than strangers. Indeed, adoptive siblings are barely similar at all. These conclusions come from massive studies employing the best instruments known to psychology. Alternative explanations that try to push the effects of the genes to zero have by now been tested and rejected.

People sometimes fear that if the genes affect the mind at all they must determine it in every detail. That is wrong, for two reasons. The first is that most effects of genes are probabilistic. If one identical twin has a trait, there is often no more than an even chance that the other twin will have it, despite having a complete genome in common (and in the case of twins raised together, most of their environment in common as well).

The second reason is that the genes' effects can vary with the environment. Although Woody Allen's fame may depend on genes that enhance a sense of humor, he once pointed out that "we live in a society that puts a big value on jokes. If I had been an Apache Indian, those guys didn't need comedians, so I'd be out of work."

Studies of the brain also show that the mind is not a blank slate. The brain, of course, has a pervasive ability to change the strengths of its connections as the result of learning and experience — if it didn't, we would all be permanent amnesiacs. But that does not mean that the structure of the brain is mostly a product of experience. The study of the brains of twins has shown that much of the variation in the amount of gray matter in the prefrontal lobes is genetically caused. And these variations are not just random differences in anatomy like fingerprints; they correlate significantly with differences in intelligence.

People born with variations in the typical brain plan can vary in the way their minds work. A study of Einstein's brain showed that he had large, unusually shaped inferior parietal lobules, which participate in spatial reasoning and intuitions about numbers. Gay men are likely to have a relatively small nucleus in the anterior hypothalamus, a nucleus known to have a role in sex differences. Convicted murderers and other violent, antisocial people are likely to have a relatively small and inactive prefrontal cortex, the part of the brain that governs decision making and inhibits impulses. These gross features of the brain are almost certainly not sculpted by information coming in from the senses. That, in turn, implies that differences in intelligence, scientific genius, sexual orientation, and impulsive violence are not entirely learned.

*

The doctrine of the blank slate had been thought to undergird the ideals of equal rights and social improvement, so it is no surprise that the discoveries undermining it have often been met with fear and loathing. Scientists challenging the doctrine have been libeled, picketed, shouted down, and subjected to searing invective.

This is not the first time in history that people have tried to ground moral principles in dubious factual assumptions. People used to ground moral values in the doctrine that Earth lay at the center of the universe and that God created mankind in his own image in a day. In both cases, informed people eventually reconciled their moral values with the facts, not just because they had to give a nod to reality, but also because the supposed connections between the facts and morals — such as the belief that the arrangement of rock and gas in space has something to do with right and wrong — were spurious to begin with.

We are now living, I think, through a similar transition. The blank slate has been widely embraced as a rationale for morality, but it is under assault from science. Yet just as the supposed foundations of morality shifted in the centuries following Galileo and Darwin, our own moral sensibilities will come to terms with the scientific findings, not just because facts are facts but because the moral credentials of the blank slate are just as spurious. Once you think through the issues, the two greatest fears of an innate human endowment can be defused.

One is the fear of inequality. Blank is blank, so if we are all blank slates, the reasoning goes, we must all be equal. But if the slate of a newborn is not blank, different babies could have different things inscribed on their slates. Individuals, sexes, classes, and races might differ innately in their talents and inclinations. The fear is that if people do turn out to be different, it would open the door to discrimination, oppression, or eugenics.

But none of this follows. For one thing, in many cases the empirical basis of the fear may be misplaced. A universal human nature does not imply that *differences* among groups are innate. Confucius could have been right when he wrote, "Men's natures are alike; it is their habits that carry them far apart."

More important, the case against bigotry is not a factual claim that people are biologically indistinguishable. It is a moral stance that condemns judging an *individual* according to the average traits of certain *groups* to which the individual belongs. Enlightened societies strive to ignore race, sex, and ethnicity in hiring, admissions, and criminal justice because the alternative is morally repugnant. Discriminating against people on the basis of race, sex, or ethnicity would be unfair, penalizing them for traits over which they have no control. It would perpetuate the injustices of the past and could rend society into hostile factions. None of these reasons depends on whether groups of people are or are not genetically indistinguishable.

Far from being conducive to discrimination, a conception of human nature is the reason we oppose it. Regardless of I.Q. or physical strength or any other trait that might vary among people, all human beings can be assumed to have certain traits in common. No one likes being enslaved. No one likes being humiliated. No one likes being treated unfairly. The revulsion we feel toward discrimination and slavery comes from a conviction that however much people vary on some traits, they do not vary on these.

A second fear of human nature comes from a reluctance to give up the age-old dream of the perfectibility of man. If we are forever saddled with fatal flaws and deadly sins, according to this fear, social reform would be a waste of time. Why try to make the world a better place if people are rotten to the core and will just foul it up no matter what you do?

But this, too, does not follow. If the mind is a complex system with many faculties, an antisocial desire is just one component among others. Some faculties may endow us with greed or lust or malice, but others may endow us with sympathy, foresight, self-respect, a desire for respect from others, and an ability to learn from experience and history. Social progress can come from pitting some of these faculties against others.

For example, suppose we are endowed with a conscience that treats certain other beings as targets of sympathy and inhibits us from harming or exploiting them. The philosopher Peter Singer of Princeton University has shown that moral improvement has proceeded for millennia because people have expanded the mental dotted line that embraces the entities considered worthy of sympathy. The circle has been poked outward from the family and village to the clan, the tribe, the nation, the race, and most recently to all of humanity. This sweeping change in sensibilities did not require a blank slate. It could have arisen from a moral gadget with a single knob or slider that adjusts the size of the circle embracing the entities whose interests we treat as comparable to our own.

Some people worry that these arguments are too fancy for the dangerous world we live in. Since data in the social sciences are never perfect, shouldn't we err on the side of caution and stick with the null hypothesis that people are blank slates? Some people think that even if we were certain that people differed genetically or harbored ignoble tendencies, we might still want to promulgate the fiction that they didn't.

This argument is based on the fallacy that the blank slate has nothing but good moral implications and a theory that admits a human nature has nothing but bad ones. In fact, the dangers go both ways. Take the most horrifying example of all, the abuse of biology by the Nazis, with its pseudoscientific nonsense about superior and inferior races. Historians agree that bitter memories of the Holocaust were the main reason that human nature became taboo in intellectual life after the Second World War.

But historians have also documented that Nazism was not the only ideologically inspired holocaust of the twentieth century. Many atrocities were committed by Marxist regimes in the name of egalitarianism, targeting people whose success was taken as evidence of their avarice. The kulaks ("bourgeois peasants") were exterminated by Lenin and Stalin in the Soviet Union. Teachers, former landlords, and "rich peasants" were humiliated, tortured, and murdered during China's Cultural Revolution. City dwellers and literate professionals were worked to death or executed during the reign of the Khmer Rouge in Cambodia.

And here is a remarkable fact: Although both Nazi and Marxist ideologies led to industrial-scale killing, *their biological and psychological theories were opposites*. Marxists had no use for the concept of race, were averse to the notion of genetic inheritance, and were hostile to the very idea of a human nature rooted in biology. Marx did not explicitly embrace the blank slate, but he was adamant that human nature has no enduring properties: "All history is nothing but a continuous transformation of human nature," he wrote. Many of his followers did embrace it. "It is on a blank page that the most beautiful poems are written," said Mao. "Only the newborn baby is spotless," ran a Khmer Rouge slogan. This philosophy led to the persecution of the successful and of those who produced more crops on their private family plots than on communal farms. And it made these regimes not just dictatorships but totalitarian dictatorships, which tried to control every aspect of life from art and education to child rearing and sex. After all, if the mind is structureless at birth and shaped by its experience, a society that wants the right kind of minds must control the experience.

None of this is meant to impugn the blank slate as an evil doctrine, any more than a belief in human nature is an evil doctrine. Both are separated by many steps from the evil acts committed under their banners, and they must be evaluated on factual grounds. But the fact that tyranny and genocide can come from an anti-innatist belief system as readily as from an innatist one does upend the common misconception that biological approaches to behavior are uniquely sinister. And the reminder that human nature is the source of our interests and needs as well as our flaws encourages us to examine claims about the mind objectively, without putting a moral thumb on either side of the scale.

OLIVER SACKS

Anybody Out There?

FROM *Natural History*

ONE OF THE FIRST books I read as a boy was H. G. Wells's 1901
fable, *The First Men in the Moon*. The two men, Cavor and Bedford,
land in a crater, apparently barren and lifeless, just before the lunar dawn; then, as the sun rises, they realize there is an atmosphere. They spot small pools and eddies of water, and then little round objects scattered on the ground. One of them, as it is
warmed by the sun, bursts open and reveals a sliver of green. ("'A
seed,' said Cavor . . . And then . . . very softly, 'Life!'") They light
a piece of paper and throw it onto the surface of the Moon. It
glows and sends up a thread of smoke, indicating that the atmosphere, though thin, is rich in oxygen and will support life as they
know it.

Here, then, was how Wells conceived the prerequisites of life: water, sunlight (a source of energy), and oxygen. "A Lunar Morning,"
the eighth chapter in his book, was my first introduction to astro-
biology.

It was apparent, even in Wells's day, that most of the planets in
our solar system were not possible homes for life. The only reasonable surrogate for the Earth was Mars, which was known to be a
solid planet of reasonable size, in stable orbit, not too distant from
the sun, and so, it was thought, having a range of surface temperatures compatible with the presence of liquid water.

But free oxygen gas — how could that occur in a planet's atmosphere? What would keep it from being mopped up by ferrous iron
and other oxygen-hungry chemicals on the surface unless, somehow, it was continuously pumped out in huge quantities, enough to

oxidize all the surface minerals and keep the atmosphere charged as well?

It was the blue-green algae, or cyanobacteria, that infused the Earth's atmosphere with oxygen, a process that took between a billion and two billion years. The fossil record shows that cyanobacteria go back three and a half billion years. Yet, amazingly, some of them still thrive today in odd corners of the world, forming strange, cushion-shaped colonies called stromatolites. It is an extraordinary experience to go to Shark Bay in western Australia, where stromatolites flourish in the hypersaline waters, to watch them slowly bubbling oxygen, and to reflect that, three billion years ago, this was how the Earth was transformed. The cyanobacteria invented photosynthesis: by capturing the energy of the sun, they were able to combine carbon dioxide (massively present in the Earth's early atmosphere) with water to create complex molecules — sugars, carbohydrates — which the bacteria could then store and tap for energy as needed. This process generated free oxygen as a byproduct — a waste product that was to determine the future course of evolution.

Although free oxygen in a planet's atmosphere would be an infallible marker of life, and one that, if present, should be readily detected in the spectra of extrasolar planets, it is not a prerequisite for life. Planets, after all, get started without free oxygen, and may remain without it all their lives. Anaerobic organisms swarmed before oxygen was available, perfectly at home in the atmosphere of the early Earth, converting nitrogen to ammonia, sulfur to hydrogen sulfide, carbon dioxide to formaldehyde, and so forth. (From formaldehyde and ammonia the bacteria could make every organic compound they needed.)

There may be planets in our solar system and elsewhere that lack an atmosphere of oxygen but are nonetheless teeming with anaerobes. And such anaerobes need not live on the surface of the planet; they could occur well below the surface, in boiling vents and sulfurous hot pots, as they do on Earth today, to say nothing of subterranean oceans and lakes. (There is thought to be such a subsurface ocean on Jupiter's moon Europa, locked beneath a shell of ice several miles thick, and its exploration is one of the astrobiological priorities of this century. Curiously, Wells, in *The*

First Men in the Moon, imagines life originating in a central sea in the middle of the Moon and then spreading outward to its inhospitable periphery.)

It is not clear whether life has to "advance" — whether evolution must take place — if there is a satisfactory status quo. Brachiopods — lampshells — for instance, have remained virtually unchanged since they first appeared in the Cambrian Period, more than 500 million years ago. But there does seem to be a drive for organisms to become more highly organized and more efficient in retaining energy, at least when environmental conditions are changing rapidly, as they were before the Cambrian. The evidence indicates that the first primitive anaerobes on Earth were prokaryotes: small, simple cells — just cytoplasm, usually bounded by a cell wall, but with little if any internal structure.

By degrees, however — and the process took place with glacial slowness — prokaryotes became more complex, acquiring internal structure, nuclei, mitochondria, and so on. The microbiologist Lynn Margulis of the University of Massachusetts, Amherst, has convincingly suggested that these complex so-called eukaryotes arose when prokaryotes began incorporating other prokaryotes within their own cells. The incorporated organisms at first became symbiotic and later came to function as essential organelles of their hosts, enabling the resultant organisms to use what was originally a noxious poison: oxygen.

Primitive as they are, prokaryotes are still highly sophisticated organisms with formidable genetic and metabolic machinery. Even the simplest ones manufacture more than five hundred proteins, and their DNA includes at least half a million base pairs. Hence it is certain that still more primitive life forms must have preceded the prokaryotes.

Perhaps, as the physicist Freeman Dyson of the Institute for Advanced Study in Princeton has suggested, there were "pro-genotes" capable of metabolizing, growing, and dividing but lacking any genetic mechanism for precise replication. And before them there must have been millions of years of purely chemical, prebiotic evolution — the synthesis, over eons, of formaldehyde and cyanide, of amino acids and peptides, of proteins and self-replicating molecules. Perhaps that chemistry took place in the minute vesicles, or globules, that develop when fluids at very different temperatures meet, as may well have happened around the boiling midocean vents of the Archaean sea.

Life as we know it is not imaginable without proteins, and proteins are built from peptides, and ultimately from amino acids. It is easy to imagine that amino acids were abundant in the early Earth, either formed as a result of lightning discharges or brought to the planet by comets and meteors.

The real problem is to get from amino acids and other simple compounds to peptides, nucleotides, proteins, and so on. It is unlikely that such delicate chemical syntheses would occur in "some warm little pond," as Darwin imagined, or on the surface of a primordial sea. Instead, they would probably require unusual conditions of heat and concentration, as well as the presence of special catalysts and energy-rich compounds to make them proceed. The biochemist Christian de Duve of Rockefeller University suggests that complex organic sulfur compounds played a crucial role in providing chemical energy, and that these compounds may have formed spontaneously early in Earth's history, perhaps in the hot, acidic, sulfurous depths of the seafloor vents (where, it is increasingly believed, life probably originated). De Duve imagines this purely chemical world as the precursor of an "RNA world," believed by many to represent the first form of self-replicating life. He thinks that the movement from one to the other was both inevitable and fast.

The two preeminent evolutionary changes in the early history of life on Earth — from prokaryote to eukaryote, from anaerobe to aerobe — took the better part of two billion years. And there then had to pass another 1,200 or 1,300 million years before life rose above the microscopic forms, and the first "higher," multicellular organisms appeared. So if the Earth's history is anything to go by, one should not expect to find any higher life on a planet that is still young. Even if extraterrestrial life has appeared, and all goes well, it could take billions of years for evolutionary processes to move it along to the multicellular stage.

Moreover, all those "stages" of evolution — including the evolution of intelligent, conscious beings from the first multicellular forms — may have happened against daunting odds. Stephen Jay Gould spoke of life as "a glorious accident"; Richard Dawkins of Oxford University likens evolution to "climbing Mount Improbable." And life, once started, is subject to vicissitudes of all kinds: from meteors and volcanic eruptions to global overheating and cooling; from dead ends in evolution to mysterious mass extinctions; and finally (if things get that far) the fateful proclivities of a species like ourselves.

We know there are microfossils in some of the Earth's most ancient rocks, rocks more than three and a half billion years old. So life must have appeared within 100 or 200 million years after the Earth had cooled off sufficiently for water to become liquid. That astonishingly rapid transformation makes one think that life may develop readily, perhaps inevitably, as soon as the right physical and chemical conditions appear.

But can one argue from a single example? Can one speak confidently of "earthlike" planets, or is the Earth physically, chemically, and geologically unique? And even if there are other "habitable" planets, what are the chances that life, with its thousands of physical and chemical coincidences and contingencies, will emerge? Life may be a one-off event.

Opinion here varies as widely as it can. The French biochemist Jacques Monod regarded life as a fantastically improbable accident, unlikely to have arisen anywhere else in the universe. In his book *Chance and Necessity*, he writes, "The universe was not pregnant with life." De Duve takes issue with this, and sees the origin of life as determined by a large number of steps, most of which must have had a "high likelihood of taking place under the prevailing conditions." Indeed, de Duve believes that there is not merely unicellular life throughout the universe but complex, intelligent life, too, on trillions of planets. How are we to align ourselves between these utterly opposite but theoretically defensible positions?

What we need, what we must have, is hard evidence of life on another planet or heavenly body. Mars is the obvious candidate: it was wet and warm there once, with lakes and hydrothermal vents and perhaps deposits of clay and iron ore. It is especially in such places that we should look, suggests Malcolm Walter, an expert on fossil bacteria that date from the Earth's earliest epochs. If the evidence shows that life once existed on Mars, we will then need to know, crucially, whether it originated there or was transported (as would have been readily possible) from the young, teeming, volcanic Earth. If we can determine that life originated independently on Mars (if Mars, for instance, once harbored DNA nucleotides different from our own), we will have made an incredible discovery — one that will alter our view of the universe and enable us to perceive it, in the words of the physicist Paul Davies, as a "biofriendly" one. It would help us to gauge the probability of finding life elsewhere instead of bombinating in a vacuum of data, caught between the poles of inevitability and uniqueness.

In just the past twenty years life has been discovered in previously unexpected places on our own planet, such as the life-rich black smokers of the ocean depths, where organisms thrive in conditions biologists would once have dismissed as utterly deadly. Life is much tougher, much more resilient, than we once thought. It now seems to me quite possible that microorganisms or their remains will be found on Mars and perhaps on some of the satellites of Jupiter and Saturn.

It seems far less likely, many orders of magnitude less likely, that we will find any evidence of higher-order, intelligent life forms, at least in our own solar system. But who knows? Given the vastness and age of the universe at large, the innumerable stars and planets it must contain, and our radical uncertainties about life's origin and evolution, the possibility cannot be ruled out. And though the rate of evolutionary and geochemical processes is incredibly slow, that of technological progress is incredibly fast. Who is to say (if humanity survives) what we may not be capable of, or discover, in the next thousand years?

For myself, since I cannot wait, I turn to science fiction on occasion — and, not least, back to my favorite Wells. Although it was written a hundred years ago, "A Lunar Morning" has the freshness of a new dawn, and it remains for me, as when I first read it, the most poetic evocation of how it may be when, finally, we encounter alien life.

STEVE SILBERMAN

The Fully Immersive Mind
of Oliver Sacks

FROM *Wired*

ONE NIGHT IN 1940, a bomb tumbled out of the sky into a garden in North London, exploding into thousands of droplets of
white-hot aluminum oxide which cascaded over the lawn. The
buckets of water that the inhabitants of the house at 37 Mapesbury
Road — two Jewish doctors and their sons — poured on the fire
only fed its chemical vehemence. Amazingly, no one was hurt, but
the brilliance of the bomb left an indelible image in the mind of
Oliver Sacks, who was seven years old the night it fell.

The thermite bomb was the second of two delivered to Mapesbury Road during the war. The first, a thousand-pound monster,
landed next door but failed to explode. Sacks remembered both
scenes vividly while writing the memoir he published last October,
Uncle Tungsten: Memories of a Chemical Boyhood. After the book was
published, however, the neurologist and author learned that his
memory had deceived him, as memories made unreliable by disorders of the brain had played tricks on the minds of the subjects of
his books. His brother Michael told him that, on the night the
thermite bomb fell, in fact, they were both away at boarding school.

"I told him, 'But I can see it *now* in my mind. Why?'" Sacks recalled last November. Michael explained that it was because their
brother David had written them a dramatic letter about the incident. Even after Sacks accepted this as fact, a visual image of the
second bomb still burned in his memory. Looking more deeply,
however, he noticed a curious difference between his memories of

the two bombs. "After the first one fell" — the bomb that didn't explode — "Michael and I went down the road at night in our pajamas, not knowing what would happen. In that memory, I can *feel* myself into the body of that little boy. And in the second memory" — the thermite bomb — "it's as if I'm seeing a brilliantly illuminated scene from a film: I cannot locate myself anywhere in the scene."

Sacks has been turning his analytical gaze inward more often these days, after four decades of studying the minds of those with such disorders as autism, Tourette's syndrome, loss of proprioception, and the sudden onset of color blindness. His tales from the borderlands of the mind, translated into twenty-one languages, have earned Sacks a worldwide readership. This month, he will be awarded the Lewis Thomas Prize by Rockefeller University, given to scientists who have made a significant achievement in literature, and his insights have been ported to a broader range of media than those of any other contemporary medical author. His 1973 book, *Awakenings,* inspired both a play by Harold Pinter and a 1990 film starring Robin Williams and Robert De Niro. Two years ago, a chapter from *An Anthropologist on Mars* also got the Hollywood treatment in a movie called *At First Sight.* His first bestseller, *The Man Who Mistook His Wife for a Hat* (published in 1985), has been turned into a one-act play, an opera, and a theatrical production in French staged by Peter Brook.

It's easy to see why directors snatch up the rights to dramatize his patients' histories. Visiting the home of an ailing music teacher, Sacks pulled the score of Schumann's *Dichterliebe* out of his bag and took a seat at the piano while the patient sang, thus discovering that the teacher's disordered mind became fluid and coherent as long as the music lasted. In the age of two-minute consultations, such stories have an obvious human charm. But less obvious are the ways that Sacks's methods have pushed against the tide of a hundred years of medical practice.

In telling the stories of his patients, Sacks transformed the genre of the clinical case report by turning it inside out. The goal of the traditional case history is to arrive at a diagnosis. For Sacks, the diagnosis is nearly beside the point — a preamble or an afterthought. Since many of the conditions chronicled by him are incurable, the force driving his tales is not the race for a remedy but the patient's striving to maintain his or her identity in a world utterly changed by the disorder. In Sacks's case histories, the hero is not the doctor or even medicine itself. His heroes are the patients who learned to tap an innate capacity for growth and adaptation amid the chaos of their disordered minds: the Touretter who became a successful surgeon, the painter who lost his color vision but found an even stronger aesthetic identity by working in black and white. Mastering new skills, these patients became even more whole, more powerfully *individual*, than when they were "well."

By restoring narrative to a central place in the practice of medicine, Sacks has regrafted his profession to its roots. Before the science of medicine thought of itself as a science, at the crux of the healing arts was an exchange of stories. The patient related a confusing odyssey of symptoms to the doctor, who interpreted the tale and recast it as a course of treatment. The compiling of detailed case histories was considered an indispensable tool of physicians from the time of Hippocrates. It fell into disrepute in the twentieth century as lab tests replaced time-consuming observation, merely "anecdotal" evidence was dismissed in favor of generalizable data, and the house call was rendered quaintly obsolete.

Our conceptions of the brain have followed a parallel course toward mechanized models of disease and healing. After the discovery in the nineteenth century that lesions in the left hemisphere of the cortex caused characteristic deficits in speech, the brain has been conceived as a complex engine built of minutely specialized parts. While the mind — the ghost in this machine — made a worthy object of study for philosophers and psychotherapists, the proper job of the neurologist was mapping the circuits that kept the thing running and figuring out which parts needed repair if the system crashed.

Until the past decade, the prevailing view of memory among neurologists hadn't evolved far beyond the ancient idea that traces of experience are embedded as literal images in the cortex — the way a signet ring would make an impression in soft wax, as Plato described. In recent years, however, advancements in cognitive neuroscience have suggested that memories unfold across multiple areas of the cortex simultaneously, like a richly interconnected network of stories rather than an archive of static files. These subliminal narratives actively shape perception and are open to retranscription — as when Sacks's brain revised the memory of his brother's letter into the image of a bomb. In his books, Sacks has long anticipated this revisioning of the mind from a passive, ghostly decoder of stimuli to an interactive, adaptive, and endlessly innovative participant in the creation of our world.

Now Sacks has turned his healing instrument on himself. In both *Uncle Tungsten* and a just-published book called *Oaxaca Journal* — an account of a fern-finding expedition in Mexico — the psyche under examination is his own.

The dynamic nature of memory was one of the things on Sacks's mind when he returned to England for a book tour last fall following the publication of *Uncle Tungsten*, his tribute to a mode of amateur scientific investigation now almost inconceivable in a world obsessed with minimizing risk. After the war, a teenage geek could walk into a chemist's shop and walk out with a supply of hydrofluoric acid. Those shops are gone now, and dull high-rises have popped up in the neighborhood around Mapesbury Road. The house where Sacks was born, occupied by his family until his father's death in 1990, was sold to the British Association of Psychotherapists. The bed in his room has been replaced by an analyst's couch.

When Sacks agreed to take me along on his expedition into what Henry James called the unvisitable past, I asked what he was most looking forward to seeing in London. "Something that I know will not be there," he replied. "The great periodic table at the Science Museum in South Kensington."

In the stratum of memories Sacks mined for *Uncle Tungsten,* the Science Museum still stands as a temple to the nineteenth-century heroic tradition in chemistry, when a boy scientist like Humphry Davy could hope to isolate new elements (he eventually discovered six) and devise experiments to overturn theories that had reigned for hundreds of years. When the museum reopened in 1945, the twelve-year-old Sacks made eager pilgrimages to its chemistry galleries, which contained flasks, balances, and retorts that had been employed by Davy, Joseph Priestley, and others in the pantheon. Michael Faraday's own chemical cabinet was on display, along with burners built by Robert Bunsen himself. But it was the sight of the periodic table that came as a revelation to Sacks.

The periodic grid of the elements first appeared in a dream to the Russian chemist Dmitri Mendeleev in 1869. Before falling asleep at his desk, the white-bearded chemist played several rounds of solitaire, and his ordering scheme may have been influenced by the arrangement of suits in the game. The table in South Kensington was an unusual one, containing not only the atomic weight, number, and symbol for each element but also samples of the elements themselves sealed in jars, bequeathed to the museum by one of Napoleon's heirs.

To the young chemist and neurologist-to-be, this grand display was an irrefutable confirmation that there was order underlying the apparent chaos of the universe, and that the human mind had been keen enough to perceive it. Now Sacks owns half a dozen T-shirts with the periodic table printed on them, along with periodic coffee mugs, tote bags, and mousepads. To spur his memories while writing the book, he filled his rooms in New York with other mnemonic triggers, including X-ray tubes, bits of amber, UV lamps, and a static electricity generator. (His unflappable personal assistant and editor, Kate Edgar, drew the line at radioactive minerals: She feared for the safety of her nine-year-old son and fretted that the hunk of pitchblende might burn a hole in the piano.)

The morning of our visit to the museum, Sacks climbed into our cab carrying what looked like a sleek gray laptop, which seemed out of character — he still writes his books by hand or on a typewriter. "It's my cushion," he explained, adding wistfully, "It's my companion." The previous day, his companion had wandered off in a cab without him. Thankfully, the driver returned it to the hotel. Sacks isn't always so lucky. "I have a great gift for losing things," he admitted.

Sacks's propensity for accidentally tossing out checks has resulted in his being banned from opening his own mail at the office. He estimates that he has lost or destroyed as many manuscripts as he's published. In 1963 he wrote a short monograph about myoclonus, the involuntary twitching of muscles that in its most severe form can be totally debilitating and in its mildest form gives rise to hiccups. He gave his only copy of the paper to a leading expert in the field, C. N. Luttrell, who committed suicide a few weeks later. Sacks was too embarrassed to ask the family for the manuscript. In 1978 another text, written on Alzheimer's disease, was given to a colleague who misplaced it while moving his office; and a briefcase containing Sacks's account of watching his first space launch (the shuttle *Atlantis,* in 1991) was stolen by a hotel thief.

"There's a metaphysical dimension to loss," Sacks observed in the cab. "I don't feel like I just left these things somewhere, I feel like there's an *annihilation field* around me — they vanish into the abyss. And once they vanish, I have to wonder if they ever existed."

He reached into the pocket of his sports jacket and produced a Japanese fan — the first of several startling objects to emerge from there, so that I came to think of the coat as having magic pockets. It was a mild winter morning, and the heat was off in the cab, but Sacks commenced fanning, explaining that he had just gotten out of a pool. Water is his native element. He swims two hours a day when he can, as he has for most of his life, scouting out pools on reading tours like a junkie cultivating reliable scores. On dry land, he is made uncomfortable by any excess of heat: He insists that the thermostats in his apartment and hotel rooms be kept at 65 degrees and has been known to show up at his office in a swimsuit. As we navigated through the London traffic, he also became anxious about time. He had to be back at the hotel in a couple of hours for a telephone session with his psychoanalyst, whom he's been seeing twice a week for thirty-five years and who addresses him as Dr. Sacks in classical Viennese fashion.

Sacks's voice is the voice of his books — precise, probing, and epigrammatic — softened by the slight anomaly that phonologists call the gliding of liquids, so that "bronze" comes out "bwonze," which gives his speech an endearing boyish quality. Age has mellowed his appearance. Back in 1961, when he was a consulting physician for the Hell's Angels in California, he set a state weightlifting record for the six-hundred-pound squat. At age sixty-eight, with his snowy beard and gold-rimmed spectacles, he still has the cherubic countenance and robust frame of a Reform rabbi who inspires a resurgence of faith in the congregation wives.

Arriving at the museum, we found the entrance dominated by a billboard advertising a new Imax theater (T-REX IN 3-D!). On the second floor, we navigated toward one of the quieter areas of the building — a gallery that seemed almost abandoned. Behind Burmese elephant weights and Chinese calipers, we found one of his old shrines intact: an exhibit devoted to the history of illumination.

Sacks was delighted and sank into a reverie. "We have a very strong feeling in my family about lighting. People take it so much for granted, but the streets were dark until about 1880," he mused in front of a display of gas mantles invented by Carl Auer von Welsbach. "Welsbach was one of my heroes. I love gas mantles — their filigree becomes incandescent with a greenish-yellow light, which is hugely nostalgic for me." Approaching a display of sodium lamps, he reached into his pocket and pulled out a spectroscope, comparing the emission spectrum of a high-pressure bulb — a muddy blur — with the distinct, saffron-yellow sodium line of an older low-pressure bulb. "Fuck these high-pressure ones!" he exulted, adding, "I have a sodium lamp in my bedroom. It's my sun."

As a boy, Sacks had explored these galleries with the same sense of freedom he felt in the natural world, seeing the periodic table as "the enchanted garden of Mendeleev." Rather than being frozen in their cases, the museum's exhibits were living manifestations of the ongoing progress of science. He would run from the museum to the library next door, where he devoured biographies of his heroes, wedding the factual underpinnings of science to the lives and personal quirks of the scientists themselves. Now the old stories awoke in him again. From behind a chunk of uranium ("You don't have a Geiger counter on you, do you?" he asked), he excavated anecdotes of Marie and Pierre Curie — the walls of their laboratory incandescent with radioactivity, and a bicycle trip they took through France between the discoveries of polonium and radium.

Once Sacks became a neurologist, he learned that recovering stories forgotten by science was crucial for his work with patients. Tourette's syndrome was considered an extremely rare and possibly fictitious disease when his *Awakenings* patients fell victim to tics and seizures caused by the experimental drug he had given them, L-dopa. He had to go back to the original reports of Gilles de la Tourette, written in the 1880s, to find useful references to the syndrome in the medical literature. It wasn't that Tourette's had been banished for nearly a century but that the people who suffered from it had become invisible to the medical establishment. Its symptoms — tics and gusts of inappropriate language, elaborate obsessions and fantasies — were hard to pinpoint in the charts and graphs of twentieth-century medicine. Only when a drug called haloperidol that could partially alleviate these symptoms came along was Tourette's "remembered" — recognized as an organic disorder, chemically and genetically based and clearly real.

By exiling the clinical anecdote to the margins of medical practice — to stories passed down in hallways from attending physician to resident — the culture of medicine had blinded itself, forgetting things it had once known. Sacks calls these knowledge gaps "scotomas," the clinical term for blind spots or shadows in the field of vision.

Even with the publication of his autobiographical books, a critical period in Sacks's background has remained in the shadows. He rarely speaks in interviews about the gap between what he calls his "chemical boyhood" and his emergence thirty years later as the author of *Awakenings*. The week we were in London, when asked if he was planning a sequel to *Uncle Tungsten,* he demurred: "I have no impulse at the moment to write a volume two. I'm not sure of the continuities between the boy who was mad for chemistry and the man I became." These transitional years are a scotoma of Sacks's own, but they were clearly important to his development as an observer of human behavior.

Our trip to London led to conversations about this period in his life. His twenties were devoted to wandering in Europe and America — often by motorcycle — with a stint in Canada in 1960, when he fought fires in British Columbia and considered joining the Canadian Air Force. That fall, he took an internship at Mount Zion Hospital in San Francisco. One of the things that drew him to the Bay Area was the presence of Thom Gunn, one of the brightest and boldest of the poets who came of age in England in the 1950s. Gunn had settled in San Francisco years earlier with his lover, an American soldier, but grew up a mile or so from the house on Mapesbury Road.

Gunn recalls the burly twenty-seven-year-old intern, who at the time went by his middle name, Wolf, telling him that he "wanted to be a writer like Freud or Darwin — someone who wrote literarily, but with scientific accuracy." Soon, typewritten pages were piling up at Gunn's door by the hundreds. "Remember when you were seventeen? When you'd start writing and keep writing through the day and night in fantastic bursts of energy? It's a wonderful madness, to produce so much. This is how Ollie has been writing books for thirty years," Gunn says. (The original manuscript of *Uncle Tungsten* was more than 2 million words long; only 5 percent of this text appeared in the final book.) Gunn enjoyed Sacks's accounts of his trips across Europe and the North American continent, hitching rides with truckers who would invite him to stash his bike in the bodies of their trucks.

Also included in the journals that Sacks gave Gunn were sharply drawn portraits of the colorful characters who populated the city's nocturnal underground. One called himself Chick O'Sanfrancisco and dressed in white leather to drive his white Harley up Polk Street; another, "Dr. Kindly," was a handsome physician and sadist who once dissected his own cat and served the meat as canapés at a party. While these sketches were "horribly accurately sarcastic," Gunn recalls, he also felt "there was a certain inhumanity to them, a rather nasty adolescent smartassness, like early Aldous Huxley — getting off on people's weaknesses. I said to him, 'You don't like people very much.'" Sacks was equally stung when someone he'd written about snapped, "Are you a human being or a tape recorder?"

After two years at Mount Zion, Sacks headed south to Los Angeles and then migrated to the Bronx in 1965. There, he met the two sets of patients who would open up his writing and his ability to empathize with his subjects: a group of migraine sufferers at Montefiore Hospital and patients at Beth Abraham who had fallen sick decades earlier with a disease that had been nearly forgotten.

At Montefiore, Sacks saw more than a thousand patients with migraine. Their symptoms fascinated him: They reported disturbances of speech, hearing, taste, touch, and vision, often seeing geometrical "auras" just before the onset of an attack, which reminded Sacks of both the mystical visions of Hildegard of Bingen and his own experiences with LSD in California. He had to go to a rare books shelf at a college library, however, to find references to migraine auras. He finally discovered rich descriptions of this phenomenon in a book by the Victorian physician Edward Liveing, which in turn contained a reference to a paper written by the astronomer John Herschel called "On Sensorial Vision." Herschel, who himself suffered from migraines, spoke of a "kaleidoscopic power" that he believed was the raw precursor to perception — the brain's assembly language, as we might say now, laid bare.

Sacks immersed himself in the neglected anecdotal literature of migraine, feeling that every one of his patients "opened out into an entire encyclopedia of neurology." In a "sudden unintended explosion" in the summer of 1967, he wrote his first book in nine days — or rather, the first incarnation of *Migraine,* which became the victim of a particularly malevolent form of the annihilation field. When he showed the book to Arnold Friedman, the chief neurologist at Montefiore, in the hope that he would write a foreword, "Friedman's face darkened," Sacks says. "He practically snatched the manuscript out of my hands and asked me how I could presume to write a book. I told him that I *had* written a book."

Friedman locked up Sacks's charts, making the clinical data inaccessible to him. "He told me that migraine was *his* subject, that it was his clinic, that I was his employee, and that any thoughts I had belonged to him. He said that if I proceeded with the book, he would see that I was fired and that I would never have another job in neurology in the United States again" — not an idle threat, as Friedman held a senior post in the American Neurological Association. "I was very easily cowed. I mentioned the situation to my father, and he told me, 'Friedman sounds like a dangerous man. You'd better lie low.' I lay low for six months, which were the most depressed, and suppressed, six months of my life." Then Sacks hatched a plan. He conspired with a janitor at Montefiore to let him into the chart room every night between one and four in the morning to transcribe all the data he could. He told Friedman he was returning to England for a vacation. "Are you going back to that book of yours?" Friedman replied ominously. The chief neurologist threatened to fire him — which he did, three weeks later, by telegram.

"I went back to London in a state of terror. Then, after ten days, I had a change of mood. I thought, 'I'm free. This man is *off* my back.'"

He redrafted the pages of *Migraine* in a week and a half and took the book to Faber and Faber, who wanted to publish it immediately. Sacks walked directly from the publisher's office for a celebratory stroll through the British Museum. "I had the most wonderful feeling, because despite internal and external forbidding, I had produced a *work,*" he told me.

A few months later Sacks returned to the United States, where he began working again at Beth Abraham with the patients he'd seen two years earlier — most of them poor, elderly Jews who had contracted "sleepy sickness" in the global encephalitis epidemic of the 1920s and then lapsed into Parkinsonian limbo. Abandoned by their families and friends, isolated from one another in the structure of the institution, they reminded Sacks of his own desolation at boarding school, where he was beaten repeatedly by a brutal headmaster.

But then L-dopa came.

He put his patients on the experimental drug. After only a few days, men and women who had been transfixed in time and space for nearly half a century, staring at the ceiling in images of living crucifixion, took steps out of their wheelchairs, danced, and sang. Then, as the limits of the drug's effectiveness became apparent, their newly awakened state was overwhelmed by tics and seizures.

A transformation occurred at Beth Abraham — not just in the patients but in Sacks. "The essential thing was that I found myself in a position of care and concern for a whole population of abandoned, forgotten, and — it first seemed — hopeless people," he recalls. "Unlike the movie of *Awakenings,* where I was portrayed living at some distance away from the hospital, I virtually lived with the patients, spending sixteen hours a day with them. I had never been in a situation of such *safe intimacy* with other human beings."

Intimacy implied responsibility, not just for the patients' well-being but for their stories, which defied the limits of traditional case reports. Sacks had transgressed the protocols of clinical practice with his L-dopa experiment: In the weeks after his first patients awakened, he abandoned the idea of a control group. Those given the drug came back into themselves, while those who took the placebo did not. Each patient responded to the drug in a unique way; they then stopped responding in ways that were also unique. "I had to try L-dopa in every patient; and I could no longer think of giving it for ninety days and then stopping — this would have been like stopping the very air they breathed," he wrote later. "No 'orthodox' presentation, in terms of numbers, series, grading of effects, et cetera, could have conveyed the historical reality of the experience."

He sent off a series of letters to the editors of the standard journals about what had happened at Beth Abraham. In his correspondence, you can hear Sacks straining at the boundaries of what could be said in the impersonal language of clinical observation: "Patient enthusiasm is likely to occur in the initial 'good' phase of drug response. Denial or minimization of adverse reactions may lead the doctor to underestimate and postpone necessary action. The requisite action, reduction, or withdrawal of the drug is likely to be strongly opposed by the patient. The third reaction is despair, seen especially during the withdrawal period." Sacks's reports were greeted with silence at first, then with sharp criticism. His experimental methods were questioned, and his accounts were criticized by a colleague at Stanford for reporting "'adverse' effects of levodopa that are at variance with most clinical reports."

The language he needed to tell his patients' stories had been pushed into the shadows, displaced by the rise of "clinimetrics" and diagnosis by machine. To communicate what happened at Beth Abraham, Sacks had to visit another nearly forgotten area of the medical literature, in which a Russian neurologist attempted to comprehend two of the strangest minds the world has ever seen.

When Sacks first paged through Aleksandr Luria's *The Mind of a Mnemonist,* he thought it was a novel. Luria had observed a patient named Sherashevsky for more than twenty-five years — a span of time during which he had seemingly forgotten almost *nothing.* One day in 1936 Luria showed him a lengthy series of nonsense syllables; in 1944 Sherashevsky could recall them perfectly. The same was true for stanzas of *The Divine Comedy* in Italian — a language he did not speak. Though Sherashevsky's memory was extraordinary, *The Mind of a Mnemonist* didn't focus on quantifying its dimensions. Instead, Luria examined the effects of having a nearly indelible memory on his patient's sense of identity. He wrote the book with obvious compassion for his subject, who drifted through a life in which his own wife and child felt less real to him than the contents of his inexhaustible memory.

Another book by Luria, *The Man with a Shattered World,* probed a mind in tragic disorder. In 1943 a Russian soldier was brought to Luria's office in Moscow. A bullet had torn into the left occipito-parietal region of the young man's brain, and scar tissue had eaten into the surrounding cortex. Waking up in a field hospital, the soldier had seen a doctor approach him and ask, "How goes it, Comrade Zasetsky?" The question made no sense to him. It was only after the doctor repeated it several times that the strange sounds resolved into words. When asked to raise his right hand, he was unable to find it. Luria asked him what town he was from, and he replied, "At home . . . there's . . . I want to write . . . but just can't."

Clearly, Zasetsky's brain had crashed. To help him, Luria needed to find a way in, conspiring with the only part of his mind that was still intact: the witnessing soul at the center of the storms in his cortex.

With tremendous effort, Luria and his assistants taught Zasetsky how to read and write again. At first, he couldn't even hold a pencil. The breakthrough came when Luria suggested that he try writing without thinking, allowing the "kinetic melody" of the movements — still remembered in his muscles — to carry his hand along. Slowly, it worked, and Zasetsky began to write out what his mind felt like from the inside. It took him all day to finish half a page, but over the next three decades, he managed to complete a diary more than three thousand pages long. *The Man with a Shattered World* was composed as a fugue for two voices: that of the doctor, with his comprehensive knowledge of neuroanatomy, and the other of his patient, who had written that he hoped one day "perhaps someone with expert knowledge of the human brain will understand my illness."

Luria's work suggested that the act of recovering one's own story was itself healing. He called the sort of writing he had done in *The Mind of a Mnemonist* and *The Man with a Shattered World* "romantic science." The two books had a profound impact on Sacks. They suggested a new form of writing that combined the clinical precision of twentieth-century neurology with both the humane observations of the great Victorian physicians and the explorations of the psyche that Freud undertook in his own case histories.

In 1972 Sacks went back to London and rented a flat within walking distance of both 37 Mapesbury Road and Hampstead Heath. When he was a boy, his mother had told him long tales about her patients — stories that were, Sacks wrote, "sometimes grim and terrifying, but always evocative of the personal qualities, the special value and valour, of the patient." His father had also regaled him with such stories. Throughout the summer, Sacks spent his mornings swimming in the ponds on the Heath and his afternoons writing the case histories that formed the heart of *Awakenings*. To understand what had happened in his patients' minds, he consulted not only neurological texts but the work of another poet who had become a friend, W. H. Auden, and the meditations on will and identity by the philosopher-mathematician Gottfried Leibniz. At night he would read the latest installments to his mother. She would interrupt him at points, saying, "That doesn't ring true." He reworked them until she said, "Now it rings true."

After *Awakenings* was published in 1973, Sacks received a letter from Thom Gunn. "The letter obsessed me for months. I carried it with me. He said that he had been 'dismayed' by my early writings and 'in despair for me as a human being.' Then he went on to say that the things which had seemed most absent in those earlier writings — empathy, affection — now seemed to be the very organizing principle of *Awakenings*. He asked me was this due to drugs, to analysis, to falling in love, or just the natural process of maturation? I wrote back and said, 'All of the above.'"

Sacks received two letters after the book's publication that were postmarked Moscow, from Luria himself. They began an intimate correspondence that lasted until Luria's death in 1977.

The "great crisis" in neuropsychology, as Sacks's Russian mentor saw it, was reconciling two modes of scientific observation. One reduces complex phenomena to their constituent parts — the way neurology had narrowed its focus from the observation of behavior to specific areas in the brain and then to individual neurons — which Luria paralleled with the evolution of chemistry, from the study of gross matter to the study of compounds, to the study of individual atoms and elements. The other mode relies on the description of phenomena and intuition to comprehend the interactivity of whole systems. Either one, he thought, was inadequate without the other.

Luria felt it was particularly crucial to reconcile these two modes when the subject of study was the brain. The left hemisphere *does* seem to function like an elaborate computer, aggregating the often imprecise or corrupt data of the senses into a panorama of the world at any given moment. But the roles of the right, and of the more recently evolved prefrontal cortex, hinge on such distinctly human qualities as the ability to plan, to imagine, to conceive of past and future, and to adapt to novel conditions. Paul Broca's studies of brain lesions in the nineteenth century, and the research that followed in their wake, had been successful at mapping the elements of the brain in isolation, increasing our understanding of how people became sick. Luria's works of romantic science, on the other hand, were *studies of how people got well*, even if they remained sick — the ways individuals managed to survive, and even thrive, despite massive disruptions to the usual order of brain business.

These studies require the neurologist to observe the patient engaged in daily life in the world outside the clinic, as Sacks has done. What we call Parkinson's disease was first noticed by the physician James Parkinson in the tics and seizures of afflicted people on the streets of London, not inside the walls of a clinic. But with the advent of mechanized models of the brain and the rage for quantifying behavior, the skills of intuitive, sharp-sighted observation that had distinguished the great minds of medicine began to wane.

In a letter to Sacks, Luria mourned, "The ability to *describe* which was so common to the great neurologists and psychiatrists of the nineteenth century . . . is almost lost now." Before Luria died, he challenged Sacks to forge a synthesis of literary and scientific observation that would do justice to the operation of the brain in the real world. Sacks undertook Luria's challenge in *The Man Who Mistook His Wife for a Hat, Seeing Voices,* and *An Anthropologist on Mars.*

In these books, Sacks provided the most vivid descriptions we have of the organic capacity for recovery and adaptation that inspired the modern age of network computing. In a book called *The Executive Brain,* Elkhonon Goldberg marvels at the parallels between the recent evolution of the higher, distributed cortical functions and the growth curve of digital networks: "Computer hardware has evolved from mainframe computers to personal computers to network personal computers . . . a gradual departure from a predominantly modular to a predominantly distributed pattern of organization reshaped the digital world." He puzzles over the fact that this "unconscious recapitulation" seems not to have been "guided by the knowledge of neuroscience." Paul Baran's original conception of a failure-resistant communications system, however — the blueprint for the Internet — was inspired by conversations with the neurobiologist Warren McCulloch, in which McCulloch described the ability of synaptic networks in brain-injured patients to route around damaged tissue (see "Founding Father," *Wired,* 9.03).

To Sacks, new models of the mind as distributed, adaptive, and endlessly creative confirm what he had already observed in his patients. His method as a physician is to collaborate with his patients to forge new pathways in their brains that restore this capacity for self-healing. He conceives of this work as an act of deep listening, attending to the subtle harmonies and disharmonies in his patients' behavior — as he wrote in *Awakenings,* "in an intuitive kinetic sympathy . . . an ever-changing, melodic, and living play of forces which can recall living beings into their own living being."

"The manner in which Oliver attends *is* the way in which he loves," observed a colleague, the neuropsychiatrist Jonathan Mueller. "The sustainedness of attention is what he does reverence with — and it's what he gives to his patients."

Sacks has raised public awareness of disorders formerly considered very rare, notably Tourette's syndrome and autism (see "The Geek Syndrome," *Wired,* 9.12). But in certain quarters, what Sacks "gives to his patients" by turning them into the subjects of best-selling books is still open to debate. A British academic and disability rights advocate named Tom Shakespeare has christened Sacks "the man who mistook his patients for a writing career." Alexander Cockburn flamed him in *The Nation* for being "in the same business as the supermarket tabloids (I MEET MONSTER FROM OUTER SPACE WITH TWO HEADS), only he is writing for the genteel classes and dresses it up a bit (I MEET MAN WHO THINKS HE'S A MONSTER WITH TWO HEADS). The bottom of it is a visit round the bin, looking at the freaks."

The Fordham University scholar Leonard Cassuto, however, points out that Sacks's case histories have precisely the opposite effect of Victorian freak shows: "Medicine killed the old-time freak show by pathologizing its exhibits. Johnny the Leopard Boy inspires no wonder and awe if you say, instead, that 'poor John is suffering from vitiligo.' Sacks is unique because he's reincarnated the freak show in precisely the same medical language that did so much to end it. People will want to stare, and Sacks is suggesting that the best way to deal with this desire is not to forbid it but rather to shape and direct it, to make the stare into a mutual look, a meeting of two worlds. Sacks uses the case history as a bridge between people with disabilities and the able-bodied majority, placing himself squarely in the middle as the link that forms the span."

Part of the way Sacks forges that link, of course, is by being visibly weird himself. For an intensely private man, he is open, even exhibitionistic, about things others might find embarrassing, such as his absent-mindedness, his ticcish idiosyncrasies, and his geeky ardor for ferns, cephalopods, and *Star Trek*. Once, while he was rushing down a crowded sidewalk in Manhattan and impatiently muttering, "Get out of my way, fucker," a man in front of him turned around and glared. "I have Tourette's syndrome, I can't help it!" Sacks said, and the man backed down. "I was shielded behind a false diagnosis," he told me, still amused by the incident.

Another aspect of Sacks's visibly odd identity is his attachment to solitude. He has never married and has not had a relationship in many years. His two most recent books, however, give the lie to the other false diagnosis frequently aimed at him — that he is asexual. In this new writing, his romance with science has become openly erotic, mining sublimated libido everywhere, even in the cryptogamic botany of cycads and the antiaircraft balloons lofted over London during the war. In *Oaxaca Journal*, he admires the "charming modesty" of ferns, their "reproductive organs . . . not thrust out flamboyantly but concealed, with a certain delicacy, on the undersides of leafy fronds." In *Uncle Tungsten*, he writes that his "first love object" was a balloon that safeguarded his neighborhood when he was ten: "I would steal over from the cricket pitch when nobody was looking and touch the gently swelling, shining fabric softly . . . It recognized and responded to my touch, I imagined, trembled (as I did) with a sort of rapture."

These polymorphous raptures extend even into the arid regions of the periodic table. After seeing the table in the Science Museum, he wrote in *Uncle Tungsten*, "I could scarcely sleep for excitement . . . I kept dreaming of the periodic table in the excited half-sleep of that night . . . The next day I could hardly wait for the museum to open." His love affair with the elements continues today in his dream life. In one recurring scenario, he is hafnium, sitting in a box at the Metropolitan Opera House alongside his companions tantalum, rhenium, osmium, iridium, platinum, gold, and tungsten. Awake, he identifies with the inert gases, a periodic group almost totally resistant to forming compounds. Also known as the noble gases, Sacks imagines them in *Uncle Tungsten* as "lonely, cut off, yearning to bond." In *Oaxaca Journal*, he refers to himself as a "singleton," which itself sounds like the name of some elementary particle.

The neurologist may have lonely nights — he calls his shyness a "disease" — but he is not without companionship. He has scores of friends and colleagues all over the world who have written books and plays, parsed the language of the deaf, alleviated the misery of devastating disorders, and one, named Patrick, who is the former captain of the starship *Enterprise*. His walls in Greenwich Village are brightened with paintings by former patients and subjects who became friends, such as the autistic artist Stephen Wiltshire and Shane Fistell, the super-Touretter in *An Anthropologist on Mars*. His familial inner circle in New York includes his assistant Kate Edgar, his analyst, his swim coach, and his archivist, Bill Morgan, who kept Allen Ginsberg's sprawling legacy in order for twenty years. (Hunting up missing missives and prodigal journals, Morgan is a human de-annihilation field.) A housekeeper comes in once a week to tame the tornado in his apartment, prepare the orange Jell-O along with the fish and tabouleh he eats every day, and generally mother him, as many of his friends seem to do.

As teddy-bearish Sacks simulacra proliferate in movies like *The Royal Tenenbaums,* he receives hundreds of letters a month — if not quite so many marriage proposals from strangers as after the movie version of *Awakenings.* A significant portion of these envelopes contain medical records from people seeking to become patients in his small private practice; many have baffling conditions and are contacting him as a physician of last resort. He still sees patients at Beth Abraham and at the Little Sisters of the Poor in Queens, for which he receives $12 per appointment. Since the publication of *Uncle Tungsten,* the daily deluge of letters, books, manuscripts, and CDs has been supplemented with specimens of mystery metals, light bulbs, and periodic tables.

While writing *Uncle Tungsten,* Sacks combed the Science Museum's archives for a photograph of the periodic table that shines in his memory, but he found only teasing near-misses taken a few years before or after the time of his pilgrimages. In the last couple of decades, the old chemistry galleries have been cleared away to make room for more "kid friendly" displays and corporate sponsorship events. The day we visited the museum, our quest for the former location of Mendeleev's garden took us to the third floor, where we came to a vacant landing. Sacks put his cushion on a step, sat down, and looked up at the white wall.

"It used to be here," he said. "That blank space is where Ollie

Sacks had his revelation of infinity and saw God. I identified Mendeleev with Moses, coming down from Sinai with the tablets of the periodic law. I visualize, and can still see as I talk, the inert gases in their huge hexagonal jars — the jars looked empty, but you *knew* they were there. There were translucent sticks of phosphorus in water, and a fist-sized lump of iridium. It must have been a pound. I adored it. There was chlorine, green and swirling in the jar. I had seen dirty bits of cesium before, but they had a lot of it; it's the only other golden metal, golden and glinting. Masurium had no atomic weight — it was not clear whether this element had been discovered or not. And crystals of iodine, all sublimed at the top of the bottle.

"That's where it was. As I close my eyes, I see the cabinet and the cubicles. Do I see a little boy standing there, or am I seeing it through the eyes of that little boy? Just yesterday. And it's fifty-five years ago."

As we got ready to leave, we paused to admire a display of photographs made to be seen through a stereoscope, the Victorian equivalent of a 3-D View-Master. (Sacks's parents had a huge collection of these images in the house on Mapesbury Road, and now he collects them himself.) In recent years, he has taken pleasure in attending meetings of groups like the New York Stereoscopic Society, where the basis of affinity isn't just a desire to mingle but a profound and exacting common interest — and one not shared by the mainstream. *Oaxaca Journal* is dedicated to the American Fern Society and to "plant hunters, birders, divers, stargazers, rock hounds, fossickers, [and] amateur naturalists the world over." Perhaps in these congregations of loners, Sacks has discovered a kind of cloud chamber — one in which even inert gases, and other rare and noble elements in the human periodic table, might find ways to bond naturally.

By beginning to write his own case history in his recent books, Sacks may be discovering what his patients and readers learned long ago: By sharing the stories of our inner lives, we recover who we are and prepare ourselves for transformation.

"I rather like having the multiple affiliations," Sacks said as we stepped out of the museum into the street. "To go from a meeting of the Fern Society to the Mineralogical Club to the Stereoscopic Society. And then I remember I'm a neurologist."

ADAM SUMMERS

Fat Heads Sink Ships

FROM *Natural History*

IN 1851 an enraged sperm whale smashed into the bow of the whaling ship *Ann Alexander,* causing it to sink in just minutes. The event resulted in a big boost in sales for the just-published *Moby-Dick,* Herman Melville's fictional account of a white sperm whale that is pursued by, but eventually sinks, the whaleship *Pequod.* (The 1820 sinking of another whaler, the *Essex,* by a sperm whale had inspired Melville's tale.)

Sperm whales (*Physeter macrocephalus*) are the largest of the toothed whales; mature males weigh more than 40 tons and stretch 50 feet from nose to fluke. But the whaling ships that sailed out of Nantucket in the mid-nineteenth century were 90 feet long and weighed nearly 250 tons. Why would a whale seek out a collision with such a ship, and — more to the point — what enabled it to survive? (To bring the question down to a more comprehensible scale, imagine your 40-pound child dashing headlong into the side of a 250-pound beached rowboat, staving a large hole in its side, then calmly picking herself up and wandering off.) The answer may be intimately connected — anatomically speaking — with the very reason the sperm whale was considered such a desirable catch.

P. macrocephalus was prized by whalers because, in addition to the oil that could be rendered from its blubber, a large quantity of higher-quality oil — spermaceti oil — could be ladled out of the enormous, thick-skinned, fiber-reinforced bulb that forms a sort of forehead. This structure, known as the spermaceti organ, has two oil-filled chambers, one of which has room for as much as 500 gallons of spermaceti oil. (The sperm whale's spermaceti organ is so big that it sometimes seems to *be* the head; on the big males, however, it plainly juts out beyond the jaw.) This organ evolved at least 20 million years ago — clearly not to sink ships. Scientists have speculated that the bulb may focus the whale's vocalizations into a tight beam, capable of sonically stunning prey, or that it may cause the sounds to resonate, thereby increasing the appeal of a whale's song to potential mates. Another theory holds that because the oil is less dense than water, the spermaceti organ is important in buoyancy control. Recently, the University of Utah researchers David Carrier and Stephen Deban, together with their undergraduate student Jason Otterstrom, proposed a pugilistic function: they think the spermaceti organ is a head-mounted boxing glove, used for combat between males.

Many other whale species also have fat-filled forehead fenders, and some have been seen using them as battering rams against their fellows. The Utah biologists observed that forehead size is closely correlated with a common measure of male-to-male aggression: sexual size dimorphism. (From fish to frogs to felines, species in which males are considerably larger than females tend to be those in which males fight for the privilege of mating; the greater the size difference between the sexes, the more competition between males.) In the species that the researchers compared, those with the most striking sexual size dimorphism were also those with the largest spermaceti organ relative to the rest of the body. A big, oil-filled forehead seems to be associated with male aggression, at least in some species. But just *how* is harder to determine.

Carrier and his group used anatomical information, including the size and shape of the spermaceti organ and of the skeleton that supports it, to build a mathematical model of imaginary collisions between jousting whales. Their goal was to see whether this organ is suited for delivering a useful broadside punch while simultaneously protecting the aggressor's noggin.

In a collision, it is not speed but rather the change in speed over time — the acceleration — that causes injury. The force on an object is the product of its mass multiplied by its acceleration. Thus, the same change in speed will exert more force on a heavier object than on a lighter one: a 3,000-pound car will be hit 100 percent harder than a 1,500-pound one when slowing from sixty miles an hour to zero upon colliding with a wall. The key to surviving a high-speed collision is to make the crash last as long as possible. Automobile designers don't aim to build cars as strong as tanks; they build cars to collapse in a controlled fashion that uses up as much of the collision's energy as possible without compromising the passenger compartment.

In accordance with this principle, an empty, blown eggshell dropped onto my counter from a height of 18 inches will not break, while a fresh egg, differing from the blown one only in mass (and in not having two tiny holes), will make a small mess. A mouse will survive a drop of several stories and land with a force of about 170 g, or 170 times the acceleration of gravity; a 10-g car crash will break human bones; and a sperm whale will suffer destructive, possibly fatal injuries at just 2 g. (The acceleration due to gravity is 32 feet per second per second.)

So how might all this pertain to a sperm whale set on slamming into a rival? First, it's helpful to take a closer look at the whale's putative battering ram. The spermaceti organ actually consists of two main chambers: a lower section called the junk, which is filled with its own oil plus baffles of connective tissue, and, atop the junk, a chamber often called the case, containing the valuable spermaceti oil. The whole organ sits on the wide upper jaw and the dished-out skull behind it. The posterior six cervical vertebrae are fused, providing a few more feet of solid, bony support.

A head-on collision between whales would be a fender-bender, with each animal's spermaceti organ cushioning the blow. But if a male could manage to thump a rival in the brisket or the chops, the outcome would be very different. A conservative estimate is that the spermaceti organ is ten times better at absorbing energy than the (relatively!) thin layer of fatty tissue that covers the rest of the whale. The model created by the Utah researchers predicts that in a broadside collision, the aggressor would experience less than 0.5 g while inflicting a dangerous, or even fatal, blow of more than 2 g. The smart thing for the intended victim to do is either move quickly out of harm's way or turn to face the danger head-on. Compared with the graceful sperm whale, nineteenth-century whaleships were slow, clumsy, and oblivious to threats from below. To the whales that sent them to the bottom of the sea, the *Essex*, the *Ann Alexander*, and the fictional *Pequod* must have seemed punch-drunk opponents just begging to be blindsided — dream targets for an angry sperm whale.

GARY TAUBES

What If It's All Been a Big Fat Lie?

FROM *The New York Times Magazine*

IF THE MEMBERS of the American medical establishment were to have a collective find-yourself-standing-naked-in-Times-Square-type nightmare, this might be it. They spend thirty years ridiculing Robert Atkins, author of the phenomenally best-selling *Dr. Atkins' Diet Revolution* and *Dr. Atkins' New Diet Revolution*, accusing the Manhattan doctor of quackery and fraud, only to discover that the unrepentant Atkins was right all along. Or maybe it's this: they find that their very own dietary recommendations — eat less fat and more carbohydrates — are the cause of the rampaging epidemic of obesity in America. Or, just possibly this: they find out both of the above are true.

When Atkins first published his *Diet Revolution* in 1972, Americans were just coming to terms with the proposition that fat — particularly the saturated fat of meat and dairy products — was the primary nutritional evil in the American diet. Atkins managed to sell millions of copies of a book promising that we would lose weight eating steak, eggs, and butter to our heart's desire, because it was the carbohydrates, the pasta, rice, bagels, and sugar, that caused obesity and even heart disease. Fat, he said, was harmless.

Atkins allowed his readers to eat "truly luxurious foods without limit," as he put it, "lobster with butter sauce, steak with béarnaise sauce . . . *bacon* cheeseburgers," but allowed no starches or refined carbohydrates, which means no sugars or anything made from flour. Atkins banned even fruit juices and permitted only a modicum of vegetables, although the latter were negotiable as the diet progressed.

Atkins was by no means the first to get rich pushing a high-fat diet that restricted carbohydrates, but he popularized it to an extent that the American Medical Association considered it a potential threat to our health. The AMA attacked Atkins's diet as a "bizarre regimen" that advocated "an unlimited intake of saturated fats and cholesterol-rich foods," and Atkins even had to defend his diet in congressional hearings.

Thirty years later, America has become weirdly polarized on the subject of weight. On the one hand, we've been told with almost religious certainty by everyone from the surgeon general on down, and we have come to believe with almost religious certainty, that obesity is caused by the excessive consumption of fat, and that if we eat less fat we will lose weight and live longer. On the other, we have the ever-resilient message of Atkins and decades' worth of best-selling diet books, including *The Zone, Sugar Busters,* and *Protein Power,* to name a few. All push some variation of what scientists would call the alternative hypothesis: it's not the fat that makes us fat but the carbohydrates, and if we eat less carbohydrates we will lose weight and live longer.

The perversity of this alternative hypothesis is that it identifies the cause of obesity as precisely those refined carbohydrates at the base of the famous Food Guide Pyramid — the pasta, rice, and bread — that we are told should be the staple of our healthy low-fat diet, and then the sugar or corn syrup in the soft drinks, fruit juices, and sports drinks that we have taken to consuming in quantity if for no other reason than that they are fat free and so appear intrinsically healthy. While the low-fat-is-good-health dogma represents reality as we have come to know it, and the government has spent hundreds of millions of dollars in research trying to prove its worth, the low-carbohydrate message has been relegated to the realm of unscientific fantasy.

Over the past five years, however, there has been a subtle shift in the scientific consensus. It used to be that even considering the possibility of the alternative hypothesis, let alone researching it, was tantamount to quackery by association. Now a small but growing minority of establishment researchers have come to take seriously what the low-carb-diet doctors have been saying all along. Walter Willett, chairman of the department of nutrition at the Harvard School of Public Health, may be the most visible proponent of testing this heretic hypothesis. Willett is the de facto spokesman of the longest-running, most comprehensive diet and health studies ever performed, which have already cost upward of $100 million and include data on nearly 300,000 individuals. Those data, says Willett, clearly contradict the low-fat-is-good-health message "and the idea that all fat is bad for you; the exclusive focus on adverse effects of fat may have contributed to the obesity epidemic."

These researchers point out that there are plenty of reasons to suggest that the low-fat-is-good-health hypothesis has now effectively failed the test of time. In particular, that we are in the midst of an obesity epidemic that started around the early 1980s and that this was coincident with the rise of the low-fat dogma. (Type 2 diabetes, the most common form of the disease, also rose significantly through this period.) They say that low-fat weight-loss diets have proved in clinical trials and real life to be dismal failures, and that on top of it all, the percentage of fat in the American diet has been decreasing for two decades. Our cholesterol levels have been declining and we have been smoking less, and yet the incidence of heart disease has not declined as would be expected. "That is very disconcerting," Willett says. "It suggests that something else bad is happening."

The science behind the alternative hypothesis can be called Endocrinology 101, which is how it's referred to by David Ludwig, a researcher at Harvard Medical School who runs the pediatric obesity clinic at Children's Hospital Boston and who prescribes his own version of a carbohydrate-restricted diet to his patients. Endocrinology 101 requires an understanding of how carbohydrates affect insulin and blood sugar and in turn fat metabolism and appetite. This is basic endocrinology, Ludwig says, which is the study of hormones, and it is still considered radical because the low-fat dietary wisdom emerged in the 1960s from researchers almost exclusively concerned with the effect of fat on cholesterol and heart disease. At the time, Endocrinology 101 was still underdeveloped, so it was ignored. Now that this science is becoming clear, it has to fight a quarter century of antifat prejudice.

The alternative hypothesis also comes with an implication that is worth considering for a moment, because it's a whopper, and it may indeed be an obstacle to its acceptance. If the alternative hypothesis is right — still a big "if" — then it strongly suggests that the ongoing epidemic of obesity in America and elsewhere is not, as we are constantly told, due simply to a collective lack of will power and a failure to exercise. Rather it occurred, as Atkins has been saying (along with Barry Sears, author of *The Zone*), because the public health authorities told us unwittingly, but with the best of intentions, to eat precisely those foods that would make us fat, and we did. We ate more fat-free carbohydrates, which in turn made us hungrier and then heavier. Put simply, if the alternative hypothesis is right, then a low-fat diet is not by definition a healthy diet. In practice, such a diet cannot help being high in carbohydrates, and that can lead to obesity and perhaps even heart disease. "For a large percentage of the population, perhaps 30 to 40 percent, low-fat diets are counterproductive," says Eleftheria Maratos-Flier, director of obesity research at Harvard's prestigious Joslin Diabetes Center. "They have the paradoxical effect of making people gain weight."

Scientists are still arguing about fat, despite a century of research, because the regulation of appetite and weight in the human body happens to be almost inconceivably complex, and the experimental tools we have to study it are still remarkably inadequate. This combination leaves researchers in an awkward position. To study the entire physiological system involves feeding real food to real human subjects for months or years on end, which is prohibitively expensive, ethically questionable (if you're trying to measure the effects of foods that might cause heart disease), and virtually impossible to do in any kind of rigorously controlled scientific manner. But if researchers seek to study something less costly and more controllable, they end up studying experimental situations so oversimplified that their results may have nothing to do with reality. This then leads to a research literature so vast that it's possible to find at least some published research to support virtually any theory. The result is a balkanized community — "splintered, very opinionated, and in many instances intransigent," says Kurt Isselbacher, a former chairman of the Food and Nutrition Board of the National Academy of Science — in which researchers seem easily convinced that their preconceived notions are correct and thoroughly uninterested in testing any other hypotheses but their own.

What's more, the number of misconceptions propagated about the most basic research can be staggering. Researchers will be suitably scientific describing the limitations of their own experiments, then will cite something as gospel truth because they read it in a magazine. The classic example is the statement heard repeatedly that 95 percent of all dieters never lose weight, and 95 percent of those who do will not keep it off. This will be correctly attributed to the University of Pennsylvania psychiatrist Albert Stunkard, but it will go unmentioned that this statement is based on 100 patients who passed through Stunkard's obesity clinic during the Eisenhower administration.

With these caveats, one of the few reasonably reliable facts about the obesity epidemic is that it started around the early 1980s. According to Katherine Flegal, an epidemiologist at the National Center for Health Statistics, the percentage of obese Americans stayed relatively constant through the 1960s and 1970s at 13 percent to 14 percent and then shot up by 8 percentage points in the 1980s. By the end of that decade, nearly one in four Americans was obese. That steep rise, which is consistent through all segments of American society and which continued unabated through the 1990s, is the singular feature of the epidemic. Any theory that tries to explain obesity in America has to account for that. Meanwhile, overweight children nearly tripled in number. And, for the first time, physicians began diagnosing Type 2 diabetes in adolescents. Type 2 diabetes often accompanies obesity. It used to be called adult-onset diabetes and now, for the obvious reason, is not.

So how did this happen? The orthodox and ubiquitous explanation is that we live in what Kelly Brownell, a Yale psychologist, has called a "toxic food environment" of cheap fatty food, large portions, pervasive food advertising, and sedentary lives. By this theory, we are at the Pavlovian mercy of the food industry, which spends nearly $10 billion a year advertising unwholesome junk food and fast food. And because these foods, especially fast food, are so filled with fat, they are both irresistible and uniquely fattening. On top of this, so the theory goes, our modern society has successfully eliminated physical activity from our daily lives. We no longer exercise or walk up stairs, nor do our children bike to school or play outside because they would prefer to play video games and watch television. And because some of us are obviously predisposed to gain weight while others are not, this explanation also has a genetic component — the thrifty gene. It suggests that storing extra calories as fat was an evolutionary advantage to our

Paleolithic ancestors, who had to survive frequent famine. We then inherited these "thrifty" genes, despite their liability in today's toxic environment.

This theory makes perfect sense and plays to our puritanical prejudice that fat, fast food, and television are innately damaging to our humanity. But there are two catches. First, to buy this logic is to accept that the copious negative reinforcement that accompanies obesity — both socially and physically — is easily overcome by the constant bombardment of food advertising and the lure of a supersize bargain meal. And second, as Flegal points out, few data exist to support any of this. Certainly none of it explains what changed so significantly to start the epidemic. Fast-food consumption, for example, continued to grow steadily through the '70s and '80s, but it did not take a sudden leap, as obesity did.

As far as exercise and physical activity go, there are no reliable data before the mid-'80s, according to William Dietz, who runs the division of nutrition and physical activity at the Centers for Disease Control; the 1990s data show obesity rates continuing to climb while exercise activity remained unchanged. This suggests the two have little in common. Dietz also acknowledged that a culture of physical exercise began in the United States in the '70s — the "leisure exercise mania," as Robert Levy, director of the National Heart, Lung, and Blood Institute, described it in 1981 — and has continued to the present day.

As for the thrifty gene, it provides the kind of evolutionary rationale for human behavior that scientists find comforting but that simply cannot be tested. In other words, if we were living through an anorexia epidemic, the experts would be discussing the equally untestable "spendthrift gene" theory, touting evolutionary advantages of losing weight effortlessly. An overweight homo erectus, they'd say, would have been easy prey for predators.

It is also undeniable, note students of Endocrinology 101, that mankind never evolved to eat a diet high in starches or sugars. "Grain products and concentrated sugars were essentially absent from human nutrition until the invention of agriculture," Ludwig says, "which was only 10,000 years ago." This is discussed frequently in the anthropology texts but is mostly absent from the obesity literature, with the prominent exception of the low-carbohydrate-diet books.

What's forgotten in the current controversy is that the low-fat dogma itself is only about twenty-five years old. Until the late '70s, the accepted wisdom was that fat and protein protected against overeating by making you sated and that carbohydrates made you fat. In *The Physiology of Taste,* for instance, an 1825 discourse considered among the most famous books ever written about food, the French gastronome Jean Anthelme Brillat-Savarin says that he could easily identify the causes of obesity after thirty years of listening to one "stout party" after another proclaiming the joys of bread, rice, and (from a "particularly stout party") potatoes. Brillat-Savarin described the roots of obesity as a natural predisposition conjuncted with the "floury and feculent substances which man makes the prime ingredients of his daily nourishment." He added that the effects of this fecula — i.e., "potatoes, grain or any kind of flour" — were seen sooner when sugar was added to the diet.

This is what my mother taught me forty years ago, backed up by the vague observation that Italians tended toward corpulence because they ate so much pasta. This observation was actually documented by Ancel Keys, a University of Minnesota physician who noted that fats "have good staying power," by which he meant they are slow to be digested and so lead to satiation, and that Italians were among the heaviest populations he had studied. According to Keys, the Neapolitans, for instance, ate only a little lean meat once or twice a week but ate bread and pasta every day for lunch and dinner. "There was no evidence of nutritional deficiency," he wrote, "but the working-class women were fat."

By the '70s, you could still find articles in the journals describing high rates of obesity in Africa and the Caribbean, where diets contained almost exclusively carbohydrates. The common thinking, wrote a former director of the Nutrition Division of the United Nations, was that the ideal diet, one that prevented obesity, snacking, and excessive sugar consumption, was a diet "with plenty of eggs, beef, mutton, chicken, butter, and well-cooked vegetables." This was the identical prescription Brillat-Savarin put forth in 1825.

It was Ancel Keys, paradoxically, who introduced the low-fat-is-good-health dogma in the '50s with his theory that dietary fat raises cholesterol levels and gives you heart disease. Over the next two decades, however, the scientific evidence supporting this theory remained stubbornly ambiguous. The case was eventually settled not by new science but by politics. It began in January 1977, when a Senate committee led by George McGovern published its *Dietary Goals for the United States,* advising that Americans significantly curb their fat intake to abate an epidemic of "killer diseases" supposedly sweeping the country. It peaked in late 1984, when the National Institutes of Health officially recommended that all Americans over the age of two eat less fat. By that time, fat had become "this greasy killer," in the memorable words of the Center for Science in the Public Interest, and the model American breakfast of eggs and bacon was well on its way to becoming a bowl of Special K with low-fat milk, a glass of orange juice, and toast, hold the butter — a dubious feast of refined carbohydrates.

In the intervening years, the NIH spent several hundred million dollars trying to demonstrate a connection between eating fat and getting heart disease, and, despite what we might think, it failed. Five major studies revealed no such link. A sixth, however, costing well over $100 million alone, concluded that reducing cholesterol by drug therapy could prevent heart disease. The NIH administrators then made a leap of faith. Basil Rifkind, who oversaw the relevant trials, described their logic this way: they had failed to demonstrate at great expense that eating less fat had any health benefits. But if a cholesterol-lowering drug could prevent heart attacks, then a low-fat, cholesterol-lowering diet should do the same. "It's an imperfect world," Rifkind told me. "The data that would be definitive is ungettable, so you do your best with what is available."

Some of the best scientists disagreed with this low-fat logic, suggesting that good science was incompatible with such leaps of faith, but they were effectively ignored. Pete Ahrens, whose Rockefeller University laboratory had done the seminal research on cholesterol metabolism, testified to McGovern's committee that everyone responds differently to low-fat diets. It was not a scientific matter of who might benefit and who might be harmed, he said, but "a betting matter." Phil Handler, then president of the National Academy of Sciences, testified in Congress to the same effect in 1980. "What right," Handler asked, "has the federal government to propose that the American people conduct a vast nutritional experiment, with themselves as subjects, on the strength of so very little evidence that it will do them any good?"

Nonetheless, once the NIH signed off on the low-fat doctrine, societal forces took over. The food industry quickly began producing thousands of reduced-fat food products to meet the new recommendations. Fat was removed from foods like cookies, chips, and yogurt. The problem was, it had to be replaced with something as tasty and pleasurable to the palate, which meant some form of sugar, often high-fructose corn syrup. Meanwhile, an entire industry emerged to create fat substitutes, of which Procter & Gamble's olestra was first. And because these reduced-fat meats, cheeses, snacks, and cookies had to compete with a few hundred thousand other food products marketed in America, the industry dedicated considerable advertising effort to reinforcing the less-fat-is-good-health message. Helping the cause was what Walter Willett calls the "huge forces" of dietitians, health organizations, consumer groups, health reporters, and even cookbook writers, all well-intended missionaries of healthful eating.

Few experts now deny that the low-fat message is radically oversimplified. If nothing else, it effectively ignores the fact that unsaturated fats, like olive oil, are relatively good for you: they tend to elevate your good cholesterol, high-density lipoprotein (HDL), and lower your bad cholesterol, low-density lipoprotein (LDL), at least in comparison to the effect of carbohydrates. While higher LDL raises your heart disease risk, higher HDL reduces it.

What this means is that even saturated fats — i.e., the bad fats — are not nearly as deleterious as you would think. True, they will elevate your bad cholesterol, but they will also elevate your good cholesterol. In other words, it's a virtual wash. As Willett explained to me, you will gain little to no health benefit by giving up milk, butter, and cheese and eating bagels instead.

But it gets even weirder than that. Foods considered more or less deadly under the low-fat dogma turn out to be comparatively benign if you actually look at their fat content. More than two-thirds of the fat in a porterhouse steak, for instance, will definitely improve your cholesterol profile (at least in comparison with the baked potato next to it); it's true that the remainder will raise your LDL, the bad stuff, but it will also boost your HDL. The same is true for lard. If you work out the numbers, you come to the surreal conclusion that you can eat lard straight from the can and conceivably reduce your risk of heart disease.

The crucial example of how the low-fat recommendations were

oversimplified is shown by the impact — potentially lethal, in fact — of low-fat diets on triglycerides, which are the component molecules of fat. By the late '6os, researchers had shown that high triglyceride levels were at least as common in heart disease patients as high LDL cholesterol and that eating a low-fat, high-carbohydrate diet would, for many people, raise their triglyceride levels, lower their HDL levels, and accentuate what Gerry Reaven, an endocrinologist at Stanford University, called Syndrome X. This is a cluster of conditions that can lead to heart disease and Type 2 diabetes.

It took Reaven a decade to convince his peers that Syndrome X was a legitimate health concern, in part because to accept its reality is to accept that low-fat diets will increase the risk of heart disease in a third of the population. "Sometimes we wish it would go away because nobody knows how to deal with it," said Robert Silverman, an NIH researcher, at a 1987 NIH conference. "High protein levels can be bad for the kidneys. High fat is bad for your heart. Now Reaven is saying not to eat high carbohydrates. We have to eat something."

Surely, everyone involved in drafting the various dietary guidelines wanted Americans simply to eat less junk food, however you define it, and eat more the way they do in Berkeley, California. But we didn't go along. Instead we ate more starches and refined carbohydrates, because calorie for calorie, these are the cheapest nutrients for the food industry to produce and they can be sold at the highest profit. It's also what we like to eat. Rare is the person under the age of fifty who doesn't prefer a cookie or heavily sweetened yogurt to a head of broccoli.

"All reformers would do well to be conscious of the law of unintended consequences," says Alan Stone, who was staff director for McGovern's Senate committee. Stone told me he had an inkling about how the food industry would respond to the new dietary goals back when the hearings were first held. An economist pulled him aside, he said, and gave him a lesson on market disincentives to healthy eating: "He said if you create a new market with a brandnew manufactured food, give it a brand-new fancy name, put a big advertising budget behind it, you can have a market all to yourself and force your competitors to catch up. You can't do that with fruits and vegetables. It's harder to differentiate an apple from an apple."

Nutrition researchers also played a role by trying to feed science into the idea that carbohydrates are the ideal nutrient. It had been known for almost a century, and considered mostly irrelevant to the etiology of obesity, that fat has nine calories per gram compared with four for carbohydrates and protein. Now it became the fail-safe position of the low-fat recommendations: reduce the densest source of calories in the diet and you will lose weight. Then in 1982 J. P. Flatt, a University of Massachusetts biochemist, published his research demonstrating that, in any normal diet, it is extremely rare for the human body to convert carbohydrates into body fat. This was then misinterpreted by the media and quite a few scientists to mean that eating carbohydrates, even to excess, could not make you fat — which is not the case, Flatt says. But the misinterpretation developed a vigorous life of its own because it resonated with the notion that fat makes you fat and carbohydrates are harmless.

As a result, the major trends in American diets since the late '70s, according to the USDA agricultural economist Judith Putnam, have been a decrease in the percentage of fat calories and a "greatly increased consumption of carbohydrates." To be precise, annual grain consumption has increased almost 60 pounds per person and caloric sweeteners (primarily high-fructose corn syrup) by 30 pounds. At the same time, we suddenly began consuming more total calories: now up to four hundred more each day since the government started recommending low-fat diets.

If these trends are correct, then the obesity epidemic can certainly be explained by Americans' eating more calories than ever — excess calories, after all, are what causes us to gain weight — and, specifically, more carbohydrates. The question is why?

The answer provided by Endocrinology 101 is that we are simply hungrier than we were in the '70s, and the reason is physiological more than psychological. In this case, the salient factor — ignored in the pursuit of fat and its effect on cholesterol — is how carbohydrates affect blood sugar and insulin. In fact, these were obvious culprits all along, which is why Atkins and the low-carb-diet doctors pounced on them early.

The primary role of insulin is to regulate blood sugar levels. After you eat carbohydrates, they will be broken down into their component sugar molecules and transported into the bloodstream. Your pancreas then secretes insulin, which shunts the blood sugar into muscles and the liver as fuel for the next few hours. This is why carbohydrates have a significant impact on insulin and fat does not. And because juvenile diabetes is caused by a lack of insulin, physicians believed since the '20s that the only evil with insulin is not having enough.

But insulin also regulates fat metabolism. We cannot store body fat without it. Think of insulin as a switch. When it's on, in the few hours after eating, you burn carbohydrates for energy and store excess calories as fat. When it's off, after the insulin has been depleted, you burn fat as fuel. So when insulin levels are low, you will burn your own fat, but not when they're high.

This is where it gets unavoidably complicated. The fatter you are, the more insulin your pancreas will pump out per meal, and the more likely you'll develop what's called "insulin resistance," which is the underlying cause of Syndrome X. In effect, your cells become insensitive to the action of insulin, so you need ever greater amounts to keep your blood sugar in check. So as you gain weight, insulin makes it easier to store fat and harder to lose it. But the insulin resistance in turn may make it harder to store fat — your weight is being kept in check, as it should be. But now the insulin resistance might prompt your pancreas to produce even more insulin, potentially starting a vicious cycle. Which comes first — the obesity, the elevated insulin, known as hyperinsulinemia, or the insulin resistance — is a chicken-and-egg problem that hasn't been resolved. One endocrinologist described this to me as "the Nobel Prize–winning question."

Insulin also profoundly affects hunger, although to what end is another point of controversy. On the one hand, insulin can indirectly cause hunger by lowering your blood sugar, but how low does blood sugar have to drop before hunger kicks in? That's unresolved. Meanwhile, insulin works in the brain to suppress hunger. The theory, as explained to me by Michael Schwartz, an endocrinologist at the University of Washington, is that insulin's ability to inhibit appetite would normally counteract its propensity to generate body fat. In other words, as you gained weight, your body would generate more insulin after every meal, and that in turn would suppress your appetite; you'd eat less and lose the weight.

Schwartz, however, can imagine a simple mechanism that would throw this "homeostatic" system off balance: if your brain were to lose its sensitivity to insulin, just as your fat and muscles do when they are flooded with it. Now the higher insulin production that comes with getting fatter would no longer compensate by suppressing your appetite because your brain would no longer register the rise in insulin. The result would be a physiologic state in which obesity is almost preordained and one in which the carbohydrate-insulin connection could play a major role. Schwartz says he believes this could indeed be happening, but research hasn't progressed far enough to prove it. "It is just a hypothesis," he says. "It still needs to be sorted out."

David Ludwig, the Harvard endocrinologist, says that it's the direct effect of insulin on blood sugar that does the trick. He notes that when diabetics get too much insulin, their blood sugar drops and they get ravenously hungry. They gain weight because they eat more, and the insulin promotes fat deposition. The same happens with lab animals. This, he says, is effectively what happens when we eat carbohydrates — in particular, sugar and starches like potatoes and rice or anything made from flour, like a slice of white bread. These are known in the jargon as high-glycemic-index carbohydrates, which means they are absorbed quickly into the blood. As a result, they cause a spike of blood sugar and a surge of insulin within minutes. The resulting rush of insulin stores the blood sugar away and a few hours later, your blood sugar is lower than it was before you ate. As Ludwig explains, your body effectively thinks it has run out of fuel, but the insulin is still high enough to prevent you from burning your own fat. The result is hunger and a craving for more carbohydrates. It's another vicious circle, and another situation ripe for obesity.

The glycemic index concept and the idea that starches can be absorbed into the blood even faster than sugar emerged in the late '70s but again had no influence on public health recommendations because of the attendant controversies. To wit: if you bought the glycemic index concept, then you had to accept that the starches we were supposed to be eating six to eleven times a day were, once swallowed, physiologically indistinguishable from sugars. This made them seem considerably less than wholesome. Rather than accept this possibility, the policymakers simply allowed sugar and corn syrup to elude the vilification that befell dietary fat. After all, they are fat free.

Sugar and corn syrup from soft drinks, juices, and the copious teas and sports drinks now supply more than 10 percent of our total calories; the '80s saw the introduction of Big Gulps and 32-

ounce cups of Coca-Cola, blasted through with sugar, but 100 percent fat free. When it comes to insulin and blood sugar, these soft drinks and fruit juices — what the scientists call "wet carbohydrates" — might indeed be worst of all. (Diet soda accounts for less than a quarter of the soda market.)

The gist of the glycemic index idea is that the longer it takes the carbohydrates to be digested, the lesser the impact on blood sugar and insulin and the healthier the food. The foods with the highest rating on the glycemic index are some simple sugars, starches, and anything made from flour. Green vegetables, beans, and whole grains cause a much slower rise in blood sugar because they have fiber, a nondigestible carbohydrate, which slows down digestion and lowers the glycemic index. Protein and fat serve the same purpose, which implies that eating fat can be beneficial, a notion that is still unacceptable. And the glycemic index concept implies that a primary cause of Syndrome X, heart disease, Type 2 diabetes, and obesity is the long-term damage caused by the repeated surges of insulin that come from eating starches and refined carbohydrates. This suggests a kind of unified field theory for these chronic diseases, but not one that coexists easily with the low-fat doctrine.

At Ludwig's pediatric obesity clinic, he has been prescribing low-glycemic-index diets to children and adolescents for five years now. He does not recommend the Atkins diet because he believes such a very low carbohydrate approach is unnecessarily restrictive; instead, he tells his patients to effectively replace refined carbohydrates and starches with vegetables, legumes, and fruit. This makes a low-glycemic-index diet consistent with dietary common sense, albeit in a higher-fat kind of way. His clinic now has a nine-month waiting list. Only recently has Ludwig managed to convince the NIH that such diets are worthy of study. His first three grant proposals were summarily rejected, which may explain why much of the relevant research has been done in Canada and Australia. In April, however, Ludwig received $1.2 million from the NIH to test his low-glycemic-index diet against a traditional low-fat, low-calorie regime. That might help resolve some of the controversy over the role of insulin in obesity, although the redoubtable Robert Atkins might get there first.

The seventy-one-year-old Atkins, a graduate of Cornell Medical School, says he first tried a very low carbohydrate diet in 1963 after reading about one in the *Journal of the American Medical Association.* He lost weight effortlessly, had his epiphany, and turned a fledgling Manhattan cardiology practice into a thriving obesity clinic. He then alienated the entire medical community by telling his readers to eat as much fat and protein as they wanted as long as they ate few to no carbohydrates. They would lose weight, he said, because they would keep their insulin down, they wouldn't be hungry, and they would have less resistance to burning their own fat. Atkins also noted that starches and sugar were harmful in any event because they raised triglyceride levels, which was a greater risk factor for heart disease than cholesterol.

Atkins's diet is both the ultimate manifestation of the alternative hypothesis as well as the battleground on which the fat-versus-carbohydrates controversy is likely to be fought scientifically over the next few years. After insisting Atkins was a quack for three decades, obesity experts are now finding it difficult to ignore the copious anecdotal evidence that his diet does just what he has claimed. Take Albert Stunkard, for instance. Stunkard has been trying to treat obesity for half a century, but he told me he had his epiphany about Atkins and maybe about obesity as well just recently when he discovered that the chief of radiology in his hospital had lost 60 pounds on Atkins's diet. "Well, apparently all the young guys in the hospital are doing it," he said. "So we decided to do a study." When I asked Stunkard if he or any of his colleagues considered testing Atkins's diet thirty years ago, he said they hadn't because they thought Atkins was "a jerk" who was just out to make money: this "turned people off, and so nobody took him seriously enough to do what we're finally doing."

In fact, when the AMA released its scathing critique of Atkins's diet in March 1973, it acknowledged that the diet probably worked but expressed little interest in why. Through the '60s, this had been a subject of considerable research, with the conclusion that Atkins-like diets were low-calorie diets in disguise; that when you cut out pasta, bread, and potatoes, you'll have a hard time eating enough meat, vegetables, and cheese to replace the calories.

That, however, raised the question of why such a low-calorie regimen would also suppress hunger, which Atkins insisted was the signature characteristic of the diet. One possibility was Endocrinology 101: that fat and protein make you sated and, lacking carbohydrates and the ensuing swings of blood sugar and insulin, you stay sated. The other possibility arose from the fact that Atkins's diet is "ketogenic." This means that insulin falls so low that you enter a state called ketosis, which is what happens during fasting and starvation. Your muscles and tissues burn body fat for energy, as does your brain in the form of fat molecules produced by the liver called ketones. Atkins saw ketosis as the obvious way to kick-start weight loss. He also liked to say that ketosis was so energizing that it was better than sex, which set him up for some ridicule. An inevitable criticism of Atkins's diet has been that ketosis is dangerous and to be avoided at all costs.

When I interviewed ketosis experts, however, they universally sided with Atkins and suggested that maybe the medical community and the media confuse ketosis with ketoacidosis, a variant of ketosis that occurs in untreated diabetics and can be fatal. "Doctors are scared of ketosis," says Richard Veech, an NIH researcher who studied medicine at Harvard and then got his doctorate at Oxford University with the Nobel laureate Hans Krebs. "They're always worried about diabetic ketoacidosis. But ketosis is a normal physiologic state. I would argue it is the normal state of man. It's not normal to have McDonald's and a delicatessen around every corner. It's normal to starve."

Simply put, ketosis is evolution's answer to the thrifty gene. We may have evolved to efficiently store fat for times of famine, says Veech, but we also evolved ketosis to efficiently live off that fat when necessary. Rather than being poison, which is how the press often refers to ketones, they make the body run more efficiently and provide a backup fuel source for the brain. Veech calls ketones "magic" and has shown that both the heart and brain run 25 percent more efficiently on ketones than on blood sugar.

The bottom line is that for the better part of thirty years Atkins insisted his diet worked and was safe, Americans apparently tried it by the tens of millions, while nutritionists, physicians, public health authorities, and anyone concerned with heart disease insisted it could kill them and expressed little or no desire to find out who was right. During that period, only two groups of U.S. researchers tested the diet, or at least published their results. In the early '70s, J. P. Flatt and Harvard's George Blackburn pioneered the "protein-sparing modified fast" to treat postsurgical patients, and they tested it on obese volunteers. Blackburn, who later became president of the American Society of Clinical Nutrition, describes his regime as "an Atkins diet without excess fat" and says he had to give it a fancy name or nobody would take him seriously. The diet was "lean meat, fish, and fowl" supplemented by vitamins and minerals. "People loved it," Blackburn recalls. "Great weight loss. We couldn't run them off with a baseball bat." Blackburn successfully treated hundreds of obese patients over the next decade and published a series of papers that were ignored. When obese New Englanders turned to appetite control drugs in the mid-'80s, he says, he let it drop. He then applied to the NIH for a grant to do a clinical trial of popular diets but was rejected.

The second trial, published in September 1980, was done at the George Washington University Medical Center. Two dozen obese volunteers agreed to follow Atkins's diet for eight weeks and lost an average of 17 pounds each, with no apparent ill effects, although their LDL cholesterol did go up. The researchers, led by John LaRosa, now president of the State University of New York Downstate Medical Center in Brooklyn, concluded that the 17-pound weight loss in eight weeks would likely have happened with any diet under "the novelty of trying something under experimental conditions" and never pursued it further.

Now researchers have finally decided that Atkins's diet and other low-carb diets have to be tested and are doing so against traditional low-calorie, low-fat diets as recommended by the American Heart Association. To explain their motivation, they inevitably tell one of two stories: some, like Stunkard, told me that someone they knew — a patient, a friend, a fellow physician — lost considerable weight on Atkins's diet and, despite all their preconceptions to the contrary, kept it off. Others say they were frustrated with their inability to help their obese patients, looked into the low-carb diets, and decided that Endocrinology 101 was compelling. "As a physician, I was trained to mock anything like the Atkins diet," says Linda Stern, an internist at the Philadelphia Veterans Administration Hospital, "but I put myself on the diet. I did great. And I thought maybe this is something I can offer my patients."

None of these studies has been financed by the NIH, and none has yet been published. But the results have been reported at conferences — by researchers at Schneider Children's Hospital on

Long Island, Duke University, the University of Cincinnati, and Stern's group at the Philadelphia V.A. Hospital. And then there's the study Stunkard had mentioned, led by Gary Foster at the University of Pennsylvania, Sam Klein, director of the Center for Human Nutrition at Washington University in St. Louis, and Jim Hill, who runs the University of Colorado Center for Human Nutrition in Denver. The results of all five studies are remarkably consistent. Subjects on some form of the Atkins diet — whether overweight adolescents on the diet for twelve weeks, as at Schneider, or obese adults averaging 295 pounds on the diet for six months, as at the Philadelphia V.A. — lost twice the weight as the subjects on the low-fat, low-calorie diets.

In all five studies, cholesterol levels improved similarly with both diets, but triglyceride levels were considerably lower with the Atkins diet. Though researchers are hesitant to agree with this, it does suggest that heart disease risk *could* actually be reduced when fat is added back into the diet and starches and refined carbohydrates are removed. "I think when this stuff gets to be recognized," Stunkard says, "it's going to really shake up a lot of thinking about obesity and metabolism."

All of this could be settled sooner rather than later, and with it, perhaps, we might have some long-awaited answers as to why we grow fat and whether it is indeed preordained by societal forces or by our choice of foods. For the first time, the NIH is now actually financing comparative studies of popular diets. Foster, Klein, and Hill, for instance, have now received more than $2.5 million from the NIH to do a five-year trial of the Atkins diet with 360 obese individuals. At Harvard, Willett, Blackburn, and Penelope Greene have money, albeit from Atkins's nonprofit foundation, to do a comparative trial as well.

Should these clinical trials also find for Atkins and his high-fat, low-carbohydrate diet, then the public health authorities may indeed have a problem on their hands. Once they took their leap of faith and settled on the low-fat dietary dogma twenty-five years ago, they left little room for contradictory evidence or a change of opinion, should such a change be necessary to keep up with the science. In this light, Sam Klein's experience is noteworthy. Klein is president-elect of the North American Association for the Study of Obesity, which suggests that he is a highly respected member of his community. And yet he described his recent experience discussing the Atkins diet at medical conferences as a learning experience. "I have been impressed," he said, "with the anger of academicians in the audience. Their response is 'How dare you even present data on the Atkins diet!'"

This hostility stems primarily from their anxiety that Americans, given a glimmer of hope about their weight, will rush off en masse to try a diet that simply seems intuitively dangerous and on which there is still no long-term data on whether it works and whether it is safe. It's a justifiable fear. In the course of my research, I have spent my mornings at my local diner, staring down at a plate of scrambled eggs and sausage, convinced that somehow, some way, they must be working to clog my arteries and do me in.

After twenty years steeped in a low-fat paradigm, I find it hard to see the nutritional world any other way. I have learned that low-fat diets fail in clinical trials and in real life, and they certainly have failed in my life. I have read the papers suggesting that twenty years of low-fat recommendations have not managed to lower the incidence of heart disease in this country and may have led instead to the steep increase in obesity and Type 2 diabetes. I have interviewed researchers whose computer models have calculated that cutting back on the saturated fats in my diet to the levels recommended by the American Heart Association would not add more than a few months to my life, if that. I have even lost considerable weight with relative ease by giving up carbohydrates on my test diet, and yet I can look down at my eggs and sausage and still imagine the imminent onset of heart disease and obesity, the latter assuredly to be caused by some bizarre rebound phenomena the likes of which science has not yet begun to describe. The fact that Atkins himself has had heart trouble recently does not ease my anxiety, despite his assurance that it is not diet-related.

This is the state of mind I imagine that mainstream nutritionists, researchers, and physicians must inevitably take to the fat-versus-carbohydrate controversy. They may come around, but the evidence will have to be exceptionally compelling. Although this kind of conversion may be happening at the moment to John Farquhar, who is a professor of health research and policy at Stanford University and has worked in this field for more than forty years. When I interviewed him in April, he explained why low-fat diets might lead to weight gain and low-carbohydrate diets might lead to weight loss, but he made me promise not to say he believed they did. He attributed the cause of the obesity epidemic to the "force-feeding of a nation." Three weeks later, after reading an article on Endocrinology 101 by David Ludwig in the *Journal of the American Medical Association,* he sent me an e-mail message asking the not-entirely-rhetorical question, "Can we get the low-fat proponents to apologize?"

BRUCE WATSON

Sounding the Alarm

FROM *Smithsonian*

A MONTH BEFORE World War II ended, a relatively unknown writer named Rachel Carson proposed an article for *Reader's Digest* about the effects of the pesticide DDT on what she called "the delicate balance of nature." The shy woman assured the editors that "it's something that really does affect everybody." They turned her down. Perhaps they felt a story about pesticides would be too depressing. Or maybe it was that DDT, then widely used in the United States, had likely saved thousands of American Marines and soldiers by killing disease-carrying insects on far-off beachheads. Carson filed the subject away and went on to write best-selling books on the wonders of the sea. A dozen years later, she decided to take up the topic again. This time would be different.

While authors and publishers like to believe that a single book can change the world, few books actually have had such an impact. Yet the day it hit bookstores forty years ago this month, Rachel Carson's *Silent Spring* fueled a vigorous public debate about the use of chemicals in our environment that has yet to be resolved. "Without this book," wrote the former vice president Al Gore in the introduction to a 1994 reprint, "the environmental movement might have been long delayed or never have developed at all." This complex, lyrical volume led not only to the banning of DDT but eventually to the formation of the U.S. Environmental Protection Agency. "After *Silent Spring*, people began to think about the chemicals they were handling, what they were doing to the environment, and what scientists weren't telling them," says Carson's biographer Linda Lear (*Rachel Carson: Witness for Nature*, 1997). "They began to question the very direction of technology."

Carson had no intention of starting a movement. Working against time following a diagnosis of cancer, she sounded her wake-up call in the name of songbirds. "If I kept silent I could never again listen to a veery's song without overwhelming self-reproach," she wrote. But in the fall of 1962, many scientists and people in the chemical industry wished she had kept silent.

Growing up in western Pennsylvania, Rachel Louise Carson, known to friends as Ray, immersed herself in nature and books, especially the sea sagas of Melville and Conrad. At the Pennsylvania College for Women in the mid-1920s, she changed her major from English to biology but retained a deep love of writing. Eventually she earned a master's degree in marine zoology from Johns Hopkins University and became a junior aquatic biologist for the U.S. Bureau of Fisheries in Washington, D.C. Her first book, *Under the Sea-Wind,* was published in 1941 and sold fewer than 2,000 copies. But it put her in contact with scientists who were beginning to ask hard questions about the fate of the Earth.

In the late 1940s, while working as publications editor for the Fish and Wildlife Service, she began her second book, *The Sea Around Us.* The literary sensation of 1951 — topping bestseller lists and winning a National Book Award — it outlined the latest science informing our understanding of the ocean. Carson almost instantly became the nation's unofficial spokesperson for the sea. "Heavens!" she wrote to a friend after winning another accolade. "Is this all about me — it is really ridiculous!" *Sea*'s success enabled her to become a full-time writer and buy a cottage on the coast of Maine, which would become a sanctuary for the rest of her life. While she would write another book about the sea, she continued to harbor nagging questions about the effect of pesticides on the land.

Dichlorodiphenyltrichloroethane (DDT) was first used as an insecticide in 1939. Just a few grains of the white powder would miraculously wipe out colonies of mosquito larvae. During World War II, B-25 bombers sprayed DDT prior to invasions in the Pacific. After the war, DDT would all but wipe out malaria in the developed world and drastically reduce it elsewhere. (The National Academy of Sciences reported in 1970 that DDT had saved more than 500 million lives from malaria.) Paul Müller, the chemist who first turned it on unsuspecting flies, won a Nobel Prize in 1948 for his work.

By the late 1950s, DDT production had nearly quintupled from World War II levels as municipal authorities took to spraying the chemical on American suburbs to eradicate tent caterpillars, gypsy moths, and the beetles that carried Dutch elm disease.

But the chemical had a disturbing characteristic: it killed indiscriminately. After finding seven dead songbirds in her yard after the area had been sprayed against mosquitoes, a Massachusetts friend of Carson's wrote a letter to the *Boston Herald* in 1958 demanding that officials "stop the spraying of poisons from the air." Carson read the letter and realized that "everything which meant most to me as a naturalist was being threatened." She decided to make DDT the subject of her next book, tentatively entitled *Man Against the Earth*.

But working on it in 1960, she was diagnosed with breast cancer and underwent a mastectomy. Subsequent radiation treatments left her nauseated and bedridden. The book she had expected to finish in a few months dragged on for four years. Finally, in June 1962, the first of a three-part excerpt from *Silent Spring* appeared in the *New Yorker* magazine.

Before the final installment hit newsstands, the Velsicol Corporation, which manufactured the pesticide chlordane (banned in 1988), threatened to sue the magazine for libel. "Everything in those articles has been checked and is true," replied the *New Yorker*'s legal counsel. "Go ahead and sue." The company never did, but the attacks had only begun. One reader wrote that Carson's work "probably reflects her Communist sympathies."

Then, in July, news broke that a supposedly harmless drug given to thousands of pregnant women in Europe for morning sickness had been determined to cause widespread birth defects. Newspapers and magazines ran photographs of babies born without arms and legs or otherwise physically deformed. "It's all of a piece," said Carson. "Thalidomide and pesticides — they represent our willingness to rush ahead and use something new without knowing what the results are going to be."

Suddenly, in a single summer, chemical science had fallen from its pedestal. By late August, reporters were asking President Kennedy if federal officials would be investigating the long-range effects of pesticides. "They already are," he answered. "I think particularly, of course, since Miss Carson's book, but they are examining the matter."

Silent Spring went on sale on September 27 and raced to the top of the *New York Times* bestseller list, where it stayed for most of the fall. By Christmas the book, which begins with Carson's fable about an idyllic countryside that teemed with wildlife until "a strange blight crept over the area and everything began to change," had sold more than 100,000 copies. In subsequent chapters, the author followed the trail of pesticides from farm to family table, provided a "Who's Who" of toxic chemicals — DDT, chlordane, malathion, parathion — and noted that pesticides accumulate in fatty tissues of organisms.

The reaction to *Silent Spring* was quick, strong, and largely negative. *Life* claimed that Carson had "overstated her case." *Time,* citing scientists' claims that insecticides were "harmless," dismissed it as an "emotional and inaccurate outburst." The chemical and food industries went after Carson aggressively. *Chemical and Engineering News,* a chemical industry trade magazine, linked Carson with "pseudo-scientists and faddists," denounced her "high-pitched sequences of anxieties," and belittled her credentials. The Nutrition Foundation mailed scathing reviews of the book to newspapers. The National Agricultural Chemicals Association launched a $250,000 campaign to refute it, and the Monsanto Corporation published a parody of Carson's opening fable, describing a world without pesticides overrun by insects and disease. In a cartoon in the November 10, 1962, issue of the *Saturday Review,* a man lamented, "I had just come to terms with fallout, and along comes Rachel Carson."

But there were voices of praise as well. The Supreme Court justice William O. Douglas called *Silent Spring* "the most important chronicle of this century for the human race."

While undergoing debilitating radiation treatments, Carson answered her critics. No civilization, she said, "can wage relentless war on life without destroying itself, and without losing the right to be called civilized." She insisted she was not against all pesticides and had never called for banning them, only for restricting their use. Public opinion wavered. Then television tipped the scales in her favor.

In April 1963, 15 million Americans watched *CBS Reports'* "The Silent Spring of Rachel Carson." "We still talk in terms of conquest," Carson said. "I think we're challenged, as mankind has never been challenged before, to prove our maturity and our mastery, not of nature but of ourselves." Her thoughtful and reserved presentation struck a chord with viewers: hundreds wrote concerned letters to Carson, CBS, the USDA, the Public Health Service, and the FDA. A month later, President Kennedy's Science Advisory Committee released its own report on pesticides, which backed Carson's thesis, criticized the government and the chemical industry, and called for "orderly reductions of persistent pesticides."

Today, despite the banning of DDT in 1972, pesticides are still widely used, and Carson still comes in for criticism. "Rachel Carson's book was a brilliant piece of writing and a seminal work, but it's clear now that she was more fearful of pesticides than was warranted," says Dennis Avery, a former senior agriculture expert with the State Department and author of *Saving the Planet with Pesticides and Plastic.* While admitting that some dangers exist to the farmers who handle concentrated amounts of pesticides, Avery maintains that the "Green Revolution" of fertilizers, pesticides, and genetically improved seeds has tripled crop yields since 1950 and saved 12 million square miles of natural habitat that otherwise would have been cleared for farmland in order to maintain the nation's food supply.

But the veteran environmentalist Barry Commoner insists that pesticides remain a significant danger to the environment and human health. "Enough is known now that we could greatly reduce and eventually eliminate the harm caused by our use of pesticides and herbicides through organic farming and integrated pest management," he says. "We are still exposed to pesticides in our diet, and not much is known about their medical consequences. Since *Silent Spring,* the only real improvement has been for the birds. Thanks to the elimination of DDT, the osprey are better off, but I don't think we are."

Silent Spring reported that chemical companies in the United States produced about 32,000 tons of pesticides in 1960. Today the EPA says that farmers, consumers, and the government use about 615,000 tons of conventional pesticides each year. (Most pesticides used today, however, are less toxic and break down faster in nature than those used forty years ago.) And, as Carson warned, insects continue to develop chemical resistance. According to the World-watch Institute, an environmental policy think tank, a higher percentage of crops in America are now lost to pests than before pesticides were first widely used. In an attempt to safeguard our food, Congress passed the Food Quality Protection Act in 1996, giving the EPA a decade to reevaluate the safety of 9,000 pesticides.

If debate over Carson's thesis continues, few doubt her impact. "Rachel Carson's legacy has less to do with pesticides than with awakening of environmental consciousness," says her biographer Lear. "She changed the way we look at nature. We now know we are a part of nature, and we can't damage it without it coming back to bite us."

In the summer of 1963, after a year of relentless controversy, Carson turned to the personal battle she knew she was bound to lose. By fall, accepting the awards that were raining down upon her, she spoke at podiums from her wheelchair. Doctors considered it a miracle that she had survived so long, yet she kept her imminent death a secret from all but her closest friends. The following spring, the public learned that the reclusive naturalist who had awakened them had died of cancer. She was fifty-six. The day after her death, Connecticut's Senator Abraham Ribicoff opened a congressional hearing on pesticides with a tribute: "Today we mourn a great lady. All mankind is in her debt."

WILLIAM SPEED WEED

The Very Best Telescope

FROM *Discover*

As TWILIGHT ENVELOPS Mount Wilson, a 5,700-foot peak near Los Angeles, Harold McAlister begins his night of stargazing by retracing the footsteps of the late astronomer Edwin Hubble. Night after night during the 1920s, Hubble headed up this same tree-lined path to scan the heavens through the 100-inch Hooker telescope — the most powerful in the world. What he saw was a bizarre universe extending far beyond the Milky Way, composed of multiple galaxies flying away from one another at breakneck speed. That discovery eventually led to the extraordinary theory about the origin of everything called the Big Bang. Now, some eighty years later, McAlister pauses along the footpath to gaze with reverence at the huge white dome protecting the famous old telescope. "That hundred-inch instrument is more important than the space telescope they named after Hubble," he says. "It's the most important telescope of the twentieth century."

Then the Georgia State University professor puts his head down and moves on. The stars are crisp above the mountain tonight — a good opportunity for him to stare at them with an entirely new kind of machine for scanning the universe. Passing behind the old observatory, he enters a long corrugated steel building marked Beam Combining Lab and arrives at the nerve center of an optical interferometer, a revolutionary device scattered across the mountaintop and composed of six conventional telescopes, 3,100 feet of light pipes, and twenty computers. It promises to transform Mount Wilson's reputation from that of keeper of a famous old telescope to the new center of cutting-edge astronomy. This is the largest of a half-dozen interferometers under construction around the world.

It is called the CHARA (Center for High Angular Resolution Astronomy) Array, and its ability to see into space with incredible detail — fifty times finer than any single-mirror telescope ever built — promises to bring the night sky into incredibly sharp focus. For example, CHARA could zoom in on an illuminated object on the moon as small as a man. "If that man were driving a car," McAlister says, "we could distinguish one headlight from another."

More important, CHARA can distinguish one star from another. That may seem odd, but most stars viewed through even the largest and newest conventional telescopes look much as they do to the naked eye — tiny dots of light, dimensionless and deceptive. Spectrographic analysis reveals that most of those pinpoints are likely to be two stars — binaries — or even more stars: Castor, for example, in the Gemini constellation, looks like a single star but is actually six balls of fire dancing around one another. Solo performers like our sun are the exception, not the rule.

Soon interferometers will help astronomers figure out why stars tend to flock together and how they behave as they age. Eventually those lessons will come back home, telling us what our sun was like in the past and exposing threats we can expect from it — giant flares, perhaps, or periods of dimming that could trigger an ice age. Interferometers will open up the heavens anew: "We'll make thousands of stellar measurements that have never been done before," McAlister says.

Interferometry is also likely to be a boon to planet hunters. If CHARA can detect individual planets around binary stars, as expected, the census of extrasolar planets will grow immensely. The more planets found, the more likely the prospects of finding planets that could support life. Searching for extraterrestrial planets could be the ultimate fulfillment of Hubble's visionary work that began here more than eight decades ago. "The hundred-inch telescope allowed us to think that the universe is broad enough and old enough for many other civilizations to have existed out there," says Robert Jastrow, director of the Mount Wilson Institute. "CHARA will restore the glory of Mount Wilson by examining stars closely for signs of ourselves."

McAlister enters a clean room in the Beam Combining Lab and slips on booties over his shoes. Inside, CHARA's associate director Theo ten Brummelaar fusses over a table of delicate optical mirrors where light waves from CHARA's six separate telescopes are combined. Tired-eyed and unshaven, ten Brummelaar has spent months struggling with complicated calibration problems, trying to get all six beams of light to meet at the same spot at the same time — the key to making interferometry work.

By contrast, the key to making better conventional telescopes is to build wider and wider mirrors. But both conventional and interferometry telescopes operate on a principle that's not exactly intuitive. When it comes to seeing detail, their ability increases as their baseline measurement increases. The baseline is the diameter across the telescope from one edge to the other; as it increases, the telescope's angular resolution increases. The surface area of the mirror is not important to sharpness and detail. Two small mirrors, one at each end of the baseline, would work just as well as a huge mirror that spans the opening. So scientists began to think about placing individual mirrors much farther apart, collecting their light, and combining the separate light waves from each telescope. The idea was popularized in the late 1800s by the Nobel Prize–winner and astronomer Albert Michelson.

Michelson took a swath of black cloth and cut two small slits in it, so that when he placed it over the 12-inch lens of his telescope, only two slits of glass showed. He pointed his masked telescope at Jupiter's moons. The moons were dimmer with the mask on because less light was collected. But Michelson discovered that only two small samples of light gave the same angular resolution as an entire 12-inch lens. And using his crude instrument, he was able to measure the diameter of the moons.

"All that matters for angular resolution is the length of the baseline," says McAlister, glancing at an architectural drawing of CHARA that hangs on the control room wall. The bird's-eye view shows six small telescopes laid out in a Y formation over the mountaintop, each feeding its collected starlight into the Beam Combining Lab via vacuum tubes. What held true for Michelson's two-holed interferometer with a 12-inch baseline, McAlister says, also holds true for CHARA — a giant six-holed interferometer with a 1,080-foot baseline. But as Theo ten Brummelaar is quick to point out, there is a catch — figuring out how to synchronize light waves from six different telescopes. It requires cutting-edge optics, super-fast computers, and new engineering invented from scratch.

*

In a conventional telescope, the curved shape of the mirror ensures that the distance the light travels from a star to the telescope's detector is the same, no matter where it hits the mirror. In Michelson's mask experiment, the curved lens sent light from each hole to the eyepiece along two paths of identical length, so the two beams arrived in sync.

With CHARA, the beams of light from the six individual telescopes must travel through a byzantine network of tubes and mirrors that lead to a computerized detector in the control room. "The separate portions of each little wave have to meet up at the detector and recognize each other as twins, as parts of the same wave," says McAlister. "If they don't arrive at exactly the same time, you see nothing." Of course, light beams from telescopes hundreds of feet apart and at different distances from the detector are not predisposed to converge at the same time. Worse, if McAlister sights a star in the western sky, its light will have an ever so slightly shorter trip to the westernmost telescope of the six than it will to the one farthest east. There are even more subtle problems to solve, too, such as tiny vibrations that can raise one scope an imperceptible sliver of an inch closer to a star than another scope.

Ten Brummelaar's challenge is to anticipate these light-path-length discrepancies and literally stall any light that arrives early. That is accomplished by "delay lines," which move mirrors up to 160 feet along rails to increase or decrease each telescope's light path. The light from each telescope travels to the combining lab through pipelines that have been pumped free of air. At the lab, each light beam hits a set of mirrors and is bumped onto a delay line, where it bounces back and forth between a mirror at one end of a rail and a mirror on a cart. A computer positions the cart at a nanometer-precise distance along the rail to stall the beam so that it is channeled to a detector at exactly the same time as the beams from the other telescopes. The farther the cart is from the mirror on the wall, the longer the delay. "It's absurd that we have to adjust the light to nanometers after it has traveled all that distance from the star," McAlister says, "but we do."

When ten Brummelaar is satisfied that all the optical equipment is in proper alignment, he and McAlister turn out the lights and step into a room in an adjoining building filled with folding tables, old office chairs, and racks of computer equipment. Taking a seat next to McAlister in front of two oversize computer monitors, ten Brummelaar taps out some commands on a keyboard. Several hundred yards away, out in the darkening night, telescope bays open. In the Beam Combining Lab, the delay lines and the movable mirrors adjust in the dark to synchronize the starlight from separate telescopes.

Tonight the astronomers are using just two of the telescopes, pointing them at large nearby stars whose diameters have already been measured using smaller interferometers. Before they can zoom in on unmeasured stars, McAlister explains, they must calibrate CHARA using stars whose dimensions are known. Ten Brummelaar aims the two telescopes, and a large white star appears, dancing on the left-hand screen. "It's dancing because of the atmosphere, like your eye sees twinkling," ten Brummelaar says. "But the picture is not the data we're after."

Instead, he and McAlister are after a complicated measurement of the "fringes," or interference patterns, of light waves from two telescopes that meet synchronously at the detector. They have programmed the system to represent the fringes as a graph, which pops up on the screen next to the dancing star. After a good deal of number crunching — to be done later, during daylight hours — the graph will show how wide that star is.

Surprisingly, astronomers using conventional telescopes have been unable to determine even the basic dimensions of the vast majority of stars, much less examine what their surfaces look like. Most of what we know about stars comes from a close analysis of just one — our sun. And even so, we know very little. Stellar astronomy, McAlister says, "has been like doing sociology while studying only one person, making broad, sweeping conclusions with an N of one. Really, we don't know: is our sun a weird Jack the Ripper anomaly or is it a nice, normal, grandmotherly star?"

The first task is to measure the diameter of stars in order to gauge their temperatures. "Temperature is the missing link in astronomy," McAlister says. "Temperature tells us what a star looks like on the inside, how it works." Once he determines the diameter of a star using CHARA, McAlister can look up its total energy output (available from conventional telescopes) and derive its temperature. It only takes a few minutes for ten Brummelaar to "get fringes" on a star and measure its diameter. Soon he will be able to zip through the firmament, measuring — each for the first time — a hundred stars a night. "It will revolutionize the field," says Charles Bailyn, chair of astronomy at Yale University. "These are the fundamental measurements that everything else relies on."

The next step in understanding stars is to look even closer — to peek at the details hidden within their diameter. When McAlister takes measurements of a star using several pairs of telescopes, he can use the data to create an image of the star surface and see whether other stars have flares and spots as our sun does. "There's no good theoretical explanation for why the sun behaves like that," McAlister says. These magnetic storms on the sun contribute to global warming here on Earth, and his extensive survey should show whether spots and flares are common and constant on other stars, whether they come and go in cycles of, say, a thousand years, or whether our sun is abnormal for having them at all.

We already know that our sun is unusual for living alone. Conventional telescopes fitted with spectrographs have determined that as many as two-thirds of stars are binaries. Even though these telescopes can "see" only one pinprick of light, the signature of a double star shows up as a cyclic Doppler shift in a spectrogram. During one-half of the stars' orbit around each other, one star of the pair is moving toward Earth in our line of sight, and its light blue-shifts in a spectrogram. The other star is moving away, and its light red-shifts. Some time later, as the stars circle each other, the first star starts moving away, red-shifted, while the other moves toward us, blue-shifted.

"Binary stars have always been called celestial vermin," McAlister jokes. That's because two stars that look like one when viewed through a conventional telescope can throw off other stellar measurements. "But CHARA," McAlister adds with wry pleasure, "is highly sensitive to vermin." He plans a large census of double stars, measuring their mass, diameter, and temperature, as well as distance of separation and the orbital motion of each pair. The data will help theoreticians figure out why most stars form in multiples, and, by contrast, why our sun formed alone. With CHARA, our understanding of stellar evolution will improve dramatically.

So will our understanding of planets. The hundred extrasolar planets discovered in recent years are all associated with single stars or widely separated binaries. Conventional planet detection uses the same spectrographic technique as conventional binary finding — a recurring Doppler shift in the light waves — and you can't look for binaries and planets at the same time. The signals get confused.

That won't happen with CHARA. What McAlister proposes is to extend his survey of binaries so that he revisits certain double stars every few months, measuring the distance between them each time. When there are no planets in a binary system, McAlister will see two stars orbiting each other smoothly, like a graceful pair of waltzers flawlessly twirling over time. But the presence of a dark planet will complicate that smooth motion like a mischievous monkey around one or both of the dancers' necks. If McAlister sees a binary star pulled in this way by something he can't see, "we'll call a press conference," he says, because they will have found a planet in a close binary system, a revolutionary find.

Greg Laughlin, an astronomer at the University of California at Santa Cruz who studies orbital dynamics, says recent calculations suggest "there's lots of room in binary systems where, theoretically, you could fit happy, stable planets." Using computer simulations based on Newton's laws of motion, researchers have found you could have a planet in a binary system orbit both stars, so long as the distance to the stars is at least three and a half times greater than the distance between them. Or you can have a planet orbit just one star, so long as it orbits at no more than one-third the distance between the two stars. "Just about every stellar system you can imagine is capable of having stable planetary orbits," says Laughlin. "Some may have habitable planet orbits."

But these are still pencil-and-paper possibilities that scientists can investigate when CHARA and other new interferometers become fully operational. "I can't tell you how long I've waited for something like this," says the University of California at Berkeley astronomer Geoff Marcy, the current king of planet finding, who has seventy extrasolar planet finds to his name. Charles Beichman, who as the chief scientist of NASA's Origins Program is charged with finding life in the cosmos, has equally high expectations for planet finding with interferometers: "If we find that binaries commonly have planets, we double the planet population of the universe. With orders of magnitude better resolution, we're now entering the golden age of astronomy."

*

As another night of tuning up CHARA ends, McAlister steps out of the Beam Combining Lab into the cool mountain air. The stars twinkling over the ghostly white dome of Hubble's grand old hundred-inch telescope are fading, and the star closest to us begins to brighten the eastern sky. For McAlister, sunrises and sunsets raise a strange thought: "Is this normal?" If binary stars have planets, and there are more binaries than solo stars, perhaps two sunrises a day is normal.

McAlister's work is full of visionary thoughts like that, but on Mount Wilson, the golden age of astronomy is unfolding without the sort of fanfare and shocking pronouncements that issued forth from here in the 1920s, when Edwin Hubble gazed into the Hooker and saw stars beyond our own galaxy for the first time. Hubble was a man for his time, full of grandeur and big statements. McAlister, by contrast, is a man of small things, of precision. The age of interferometry is not about seeing farther; it's about seeing more clearly. McAlister spends his nights delaying light waves with mirrors that must be positioned to the millionth of an inch. The golden age of astronomy is in the details.

SCOTT WEIDENSAUL

Raising the Dead

FROM *Audubon*

THE SMALL ANIMAL is wrinkled and gray, its forelegs curled against its chest in an unintentionally protective position and a long, open incision running the length of its stomach. Stored in a clear jar of alcohol, it hardly looks like the stuff of high-tech science and acrimonious debate. This young thylacine, the marsupial "tiger" of Tasmania, was taken from its mother's pouch in 1866 — at first a curiosity from a weird and newly settled land, later a pitiable relic of a species recklessly driven to extinction.

But now, 136 years after its death and 66 years after its species was declared extinct, the preserved baby sits at the junction of molecular biology, conservation ethics, and endangered species politics — and also at the locus of humanity's guilt and hopes in dealing with the natural world. That's a lot to pin to a dead creature you could easily cup in two hands, but ever since 1999, when the Australian Museum in Sydney announced its intention to clone a living thylacine from this pickled specimen, the reaction, both pro and con, has been surprisingly fierce. Critics have lambasted it as science fiction that will drain money from more important work; proponents see it as a way to mitigate a grievous wrong committed against the planet while burnishing Australia's languishing scientific reputation.

The man at the center of the storm is Don Colgan, a soft-spoken evolutionary biologist who heads the museum's cloning team. He admits the odds of ever producing a live thylacine are long, but his group has already managed to extract unusually good DNA from the preserved baby. Several other cloning projects around the world have focused on long-extinct species, but only the thylacine team has posted any significant success to date. If Colgan succeeds where so many have predicted failure, he will not only have beaten his own odds but restored — at least in facsimile — one of the natural world's most unusual masterpieces.

To understand why the museum's project has attracted such attention, it's important to understand the significance of the thylacine from both a biological and a cultural perspective. Australia's unique, marsupial-dominated fauna have long been a textbook example of convergent evolution, and in this the thylacine is Exhibit A. Though more closely related to kangaroos and opossums, it was molded by the demands of its predatory lifestyle into a close analog of the wolf — a lean hunter with a short brownish coat, a stiff tail, and more than a dozen dark stripes along its back and hindquarters. At roughly 65 pounds, the thylacine was the largest of the carnivorous marsupials and an accomplished hunter of wallabies and other grazing species in the coastal scrublands, eucalyptus forests, and alpine meadows of Tasmania — "Tassie" to most Australians.

English settlers in the early nineteenth century — the first Europeans to have any close dealings with the thylacine — variously dubbed it the marsupial wolf, a native cat, or even a hyena. But the name that stuck was Tasmanian tiger, even though there was nothing remotely feline about the animal. (The name *thylacine,* coined by scientists, comes from the Greek words for "pouched dog.") Regardless of what they called it, sheep farmers accused the thylacine of attacking their flocks, and after a century of trapping, shooting, and poisoning, the last known thylacine died in a zoo in 1936 — ironically, less than two months after the Tasmanian government extended legal protection to the species.

But Tassie has been reluctant to let go of its tiger. The beast has become an icon — stylized on automobile license plates, gracing the label of the state's bestselling beer, adopted as the symbol of everything from a regional television network to local sports teams. Of the animal itself, though, there is nothing but specimens locked up in museum cases or gathering dust on shelves. Most consist of stuffed skins or skeletal material, but a fair number are pouch babies, the nearly hairless neonates too young to be out on their own. The usual method for preserving such soft-tissue specimens is to submerge them in formaldehyde, but the Australian Museum's now-famous baby is embalmed in alcohol. That makes all the difference, Don Colgan explained as he guided me through the warren of narrow corridors and cluttered labs in the museum's research wing.

Formaldehyde is hard on DNA's double helix of nucleic acids, Colgan said, but alcohol is much gentler — meaning that it has been possible to extract relatively high-quality DNA from several tiny samples of the thylacine's heart, liver, and muscle. Colgan pulled a large X ray from a file and held it up to the light, showing me five lines a couple of inches long and heavily crossbarred with light and dark bands — the DNA, treated with radioactive nucleotides and photographed. With his index finger, he indicated one of the blurry streaks.

"The DNA represented on this line would be about forty copies of every gene in the thylacine genome," Colgan said. "That doesn't sound a lot, but it was extracted from probably a match-head-size piece of tissue." Scientists measure DNA fragments by the number of base pairs they contain — the rungs that form the twisted ladder of a DNA molecule. This thylacine's DNA contains 1,200 to 2,000 base pairs per section — badly fragmented when compared with samples from living organisms but ten times better than is normal with ancient DNA.

Given the surprisingly good results, the museum's team is preparing to create a bacterial genomic library by inserting the thylacine genes into bacteria, which will allow the scientists to grow as much of the genetic material as they need. Beyond that, they hope to have the Tasmanian tiger genes sequenced — their genomic code "read" in its proper order. "It's like putting a jigsaw puzzle together; if you have large pieces covering the same area as a number of very small pieces, obviously it's a lot easier with the large pieces," Colgan said.

In the near term, genomic sequencing promises several scientific payoffs, he said, including genetic comparisons with other marsupials. But it is the prospect of cloning — creating a living, breathing thylacine — that raises the greatest expectations and presents the most serious challenges. In normal cloning, the nucleus of a host egg cell is removed and another living cell, containing the DNA blueprint for the organism being cloned, is inserted. (Because each clone is an exact genetic copy of the parent specimen, one must have multiple specimens of both sexes and a variety of family lineages to create a reproducing population. Colgan noted that there are hundreds of thylacine specimens in the world's museums, many of which may provide equally usable DNA.)

The trouble is that with ancient DNA, there is no living cell nucleus to serve as a starting point, so unless someone invents a way to tease a dead cell back to life, Colgan and his team must use what he calls "the brute force approach." This entails reading what they can of the thylacine genome and filling in any gaps with DNA from other marsupials, then creating artificial chromosomes, packaging them in artificial membranes, and inserting them into an egg from a closely related species. The Tasmanian devil and the numbat, the latter a small termite hunter, are prime candidates for supplying both missing DNA and a host egg.

If that sounds a bit like science fiction, you're close. Creating an artificial chromosome would require assembling roughly 50,000 pieces of DNA, each containing about 2,000 base pairs, in exactly the correct order for each of the thylacine's chromosomes, which may number between thirty and eighty — a task currently beyond anyone's capability and likely to remain so for the foreseeable future. What's more, scientists are learning that even with conventional cloning, random errors seem to creep into the genetic code and create all manner of flaws in the cloned animal, from obesity and heart defects to immune system malfunctions.

The quality of the DNA extracted from the preserved pup makes at least the sequencing part of the job easier, Colgan said, and he is confident that decoding the tiger's genome is simply a matter of time. He's exploring partnerships with large sequencing companies, which could run through the thylacine's 3.5 billion base pairs in a few years. But the price tag would be high; while the initial work has been relatively inexpensive, just sequencing the genomes of the thylacine and potential host species may run to $15 million each. (Funding so far has come from the Australian Museum, the government of New South Wales, and a private trust set up by two brothers specifically to support the cloning effort.)

Even assuming his team can sequence the tiger's complete genetic library, Colgan pegs the chances of cloning a live thylacine at only 30 percent over two hundred years. "If the project ever does come to fruition and we have a bounding baby thylacine, it'll be because of advances we simply can't conceive of at the moment," he said.

Another scientist who is using cloning technology, however, says the roadblocks facing the Australian Museum are almost insurmountable. Alan Trounson heads a small team at the Monash Institute for Reproduction and Development, in Melbourne, that's working to clone the critically endangered northern hairy-nosed wombat. Trounson told me the thylacine project faces two barriers, one technical and the other biological. First, he said, barring advances in the synthesis of artificial chromosomes, the alcohol that preserved the DNA so well for sequencing also negates the chance of ever using those cells directly for cloning, since there is no way to rehydrate them once the alcohol is removed and the cellular structure collapses.

"I am very supportive of their looking at the DNA sequence and perhaps comparing it to the DNA of existing Australian mammals — I think that could be very valuable," Trounson said. "But if they're talking about cloning a living animal, there's no way they can use cells that have been fixed in alcohol. I know what alcohol does to cells, and I can't for the life of me see how they think they're going to do it."

But even if they progress beyond that, there's a bigger hurdle remaining.

"If it were technically feasible, and we had cells that were theoretically alive — that is, the nucleus was whole and functional — then what do you put it in?" asked Trounson. He dismisses the idea of using Tasmanian devils or numbats as surrogates, noting that the thylacine split off from other marsupials an estimated 30 million years ago, leaving it with no near cousins — unlike the wombat, which has a very close relative to serve as a surrogate mother.

Others in the Australian conservation community look askance at the thylacine project, not on scientific grounds but for reasons of priority and ethics. I left Sydney and flew the six hundred miles southwest to Hobart, the capital of Tasmania, to meet with Michael Lynch, the executive director of the Tasmanian Conservation Trust. He didn't mince words when asked for his opinion.

"It just seems so loopy," said Lynch, a bluff, open-faced guy in his late fifties with a full head of curly, gray-white hair. "It's a bit like boys with their toys, like men with their big guns and their bulldozers. It's about science for the sake of science.

"Like the United States, we've got thousands of species on the endangered list," Lynch continued, "and we are just so poor in terms of the ability of the commonwealth and the states to fund threatened-species-recovery programs. If somebody gave the Tasmanian Conservation Trust that amount of money, I could run fifty recovery programs, and with the people I could have at my disposal, I could guarantee a bloody good success rate. And these are species that are here now that are being threatened."

A third batch of critics contend that trying to clone extinct species is tantamount to playing God, an act of almost Frankensteinesque hubris. Others, however, see the issue in a very different light. A conference of bioethicists, scientists, scholars, and native tribes convened in New Zealand in 1999 to consider the plan to clone an extinct crow-size bird called the huia eventually concluded that the project was a moral imperative, an act of "restorative justice."

Perhaps the ultimate irony about the strange and continuing tale of the thylacine is that while one group argues about the propriety of bringing it back from extinction, other experts contend that reports of its demise are premature. More than a few people think the animal survives — not in a metaphysical sense in test tubes in Sydney but in the forests, button-grass plains, and fern gullies of northern Tasmania, where a steady trickle of sightings keeps alive the hope that tigers still linger in their mountain fastness. And as I discovered, after leaving Sydney for two weeks of bush walking through Tasmania, just the idea of surviving thylacines adds to the luster of an already spectacular landscape.

It was a cool, breezy day, with random shafts of sunlight illuminating the thick carpets of green moss that cloaked the ground and tree trunks of the rain forest. I was hiking high in the Great Western Tiers, a range of rugged mountains that frame Tasmania's Central Plateau — a steep escarpment covered with old-growth eucalyptus, dense stands of deciduous beeches, and tree ferns with fronds like ostrich feathers.

I was here in part because of a conversation I'd had some days earlier with Eric Guiler, a retired zoology professor from the University of Tasmania who has been trailing the thylacine's fading track for nearly fifty years. Now in his eighties but still an active field researcher, Guiler is the undisputed expert on the Tassie tiger, having mounted several major expeditions in the 1950s and 1960s to search for the animal, and he has written extensively on the subject. He has been publicly critical of the museum's cloning project in the past — at one point, he called its proponents "bloody idiots" — but when I asked him about it, he simply shrugged it off, saying he doesn't know enough about molecular biology to evaluate its merits.

To Guiler, the bigger question is whether the tiger still haunts wild Tasmania. He is confident the thylacine survived long past its official extinction date of 1936 — he interviewed old opossum trappers who admitted snaring tigers through at least the 1950s, though they disposed of the carcasses for fear of prosecution. Reports of live thylacines come in to the state wildlife department on a monthly basis — many of them obvious cases of mistaken identity or wishful thinking but others hard to dismiss, like that from a Tasmanian Parks and Wildlife ranger who said he saw a tiger in the beam of his spotlight in 1982.

Most of the reports, Guiler told me, center on two areas — the rugged northwestern corner of the island, including the Western Tiers, and northeastern Tasmania, where he thinks it is significant that the number of alleged sightings has dwindled to a handful in recent years — a sign, perhaps, that the last tigers are winking out there. Yet when I asked what should be done to nurture any survivors, his frustration flashed to the surface.

"There's nothing you *can* do," Guiler said flatly. "You've got an animal about whose behavior we know virtually nothing, about whose movements we know nothing, about whose general physiology we know nothing. We know so little about it, it's impossible to devise a management plan. We don't even know where the damned things are living. How do you devise a management plan for something like that?"

In the end, although he remains defiantly hopeful, even Guiler can't be sure the Tasmanian tiger still moves, wraithlike, in the shade of the swamp gums and stringybarks, sending the wallabies scattering in fear, as it did for millions of years.

*

After several hours of climbing through dark forest, rising higher and higher toward the edge of the escarpment, the trail I was on broke into the open. The small river I'd been following went crashing down a sheer, 400-foot cliff broken in the middle of its drop by a single ledge; with a little careful climbing I was able to scramble out that shelf to sit beside the rushing waterfall. The view down the valley was breathtaking: bare talus fields of orangish dolerite rock littered the sides of the high peaks, below which grew a solid blanket of eucalyptus and beech forest. The long valley, gouged into a wide U by ice age glaciers, faded to blue in the distance, with rank after rank of mountains, hazy and indistinct, on either hand.

Once this was tiger country — and perhaps it still is, if Guiler and the other die-hard optimists are right, though I find it hard to believe that in all these years, no thylacine has stepped in front of a car, like the hundreds of wallabies and opossums that die each night on Tassie's winding roads. And so let us say for the moment that the thylacine exists only as museum specimens. If, by a miracle of molecular science, Don Colgan and his colleagues succeed in creating a living, breathing animal, what will they have? One individual does not a species make, and even with many old specimens to mine for DNA, building a population of a few dozen unique tigers would be the labor of decades and many more millions of dollars.

Today the thylacine is both a wraith and a warning, clad in a striped coat. Clone one, the skeptics say, and it becomes a novelty; clone a couple of dozen, and they become a conservation headache — a population that needs a home, and management, and oversight, just another endangered species in a world already parceling out limited resources to so many others. It might also deaden the public's worry over critically rare organisms by planting the comfortable but incorrect notion that there's a scientific quick fix even for something as permanent as extinction.

There is also a less tangible question, one that strikes me hard as I sit in the bright sun and look out over Tassie's beautiful mountains. How much of what made a thylacine a thylacine — the way it moved, the way it hunted, the way it integrated itself with its prey and its environment — was a matter of hard-wiring in the genes, and how much was a matter of cultural transmission from parent to offspring, in links that spanned 30 million years but are now irrevocably broken? How much came from within — and might be reconstituted from strings of synthesized nucleic acids wrapped in an artificial membrane — and how much came from without, from the tiger's connection with the land itself? What relation would a cloned tiger have to the animal that met the first Aborigines 40,000 years ago or watched with hungry eyes as the tall sailing ships unloaded their bleating cargo?

Yet despite the doubts, the roadblocks, the uncertainties, and the astronomical price tag, it's hard to argue with another feeling I had that day, and on many days in the forests of Tasmania — the almost visceral absence of a capstone predator, which even the project's critics acknowledge may provide the noblest motive for the drive to resurrect the thylacine. It made me think of a comment Michael Lynch had made, his eyes staring off somewhere over my shoulder as he spoke.

"It seems to me that one of the things in play here is a recognition that we, the human race, caused these animals to go to extinction," he said. "And at some level we're saying, if we can find one out there in the wild, or if we can clone one, we're assuaging that guilt somehow." Perhaps we need the thylacine more than the thylacine, sleeping in its jar of alcohol, needs us.

STEVEN WEINBERG

The Truth About Missile Defense

FROM *The New York Review of Books*

1.

ON DECEMBER 13, 2001, President Bush announced that in six months the United States would withdraw from the 1972 ABM treaty, a treaty that limits the testing and prohibits the deployment of any national missile defense system by Russia or the United States. The stated reason for this decision was that the United States needs to develop a system that would protect us from attack by intercontinental ballistic missiles launched by terrorists or by a so-called rogue state. The United States has not yet withdrawn from the treaty; this is the formal six months' advance notice that is required by the treaty, and the president could still decide not to withdraw, but it is hard to imagine that anything could happen before June 2002 that would change his mind.

The arguments by scientists and members of Congress that the United States could continue an active program of developing and testing missile defense systems without abrogating the ABM treaty now seem moot. But the issue of whether to actually develop and deploy a national missile defense system is not moot, and will not be settled even after the treaty is abrogated. Requests for missile defense funding will come up again in Congress in mid-2002, and in subsequent years. We can anticipate a continuing national debate about whether the United States should seek to develop and deploy a national system of defense against intercontinental ballistic missiles.

Few of the arguments in this debate will be new. Indeed, it is hard to remember a time when the United States has not been arguing about a national missile defense program. Almost half a century ago, in the Eisenhower administration, the Army proposed to convert the old Nike antiaircraft system to an antimissile system called Nike Zeus, which would send radar-guided nuclear-armed rockets to intercept Soviet warheads as they plunged through the atmosphere toward U.S. cities. It had obvious failings: the nuclear blasts from successful interceptions could put our radars out of action, and the stock of interceptor missiles could be exhausted if the enemy missiles carried several light decoys along with each warhead.

In the Kennedy administration the Nike Zeus plan was upgraded to a two-tier project called Nike X. Long-range nuclear-armed missiles called Spartans would attempt to intercept Soviet missiles while they were still coasting above the Earth's atmosphere; short-range Sprint missiles would then deal in the atmosphere with those warheads that had survived the Spartan attack. As a member of the JASON group of defense consultants, I worked in the 1960s on the problem of discriminating decoys from warheads and learned how difficult it is. Like others before me, I gradually also became influenced by a powerful argument against deploying any missile defense system: that in the conditions of the times it would simply induce the Soviets to increase their offensive intercontinental missile forces, leaving us worse off than before.

Despite such arguments, the Johnson administration came under powerful political pressure to go ahead with some sort of missile defense. In 1967 Defense Secretary Robert McNamara gave a remarkable speech in which he explained all the reasons against deploying a national missile defense, then concluded that the Johnson administration would go ahead anyway with a limited antimissile system, now to be called Sentinel, which would protect our cities only from attack either by accident or by what was then considered to be a rogue state, China.

To everyone's surprise, the most effective opposition to the Sentinel system did not come from experts who criticized its effectiveness or worried about arms control, but rather from citizens who simply did not want nuclear-armed defensive missiles in their neighborhoods. In response to this opposition, the Nixon administration moved the proposed Sprint missile sites away from cities and renamed the system Safeguard. Its declared purpose was now to defend our offensive missile silos instead of our cities against a missile attack. This was intended to defuse worries about strategic stability — protecting our missile silos would not make it necessary for the Soviets to increase their forces in order to maintain their ability to retaliate for a U.S. first strike. And by protecting our own offensive missiles, Safeguard would reduce any incentive that we might have to launch missiles in a crisis. As explained by Defense Secretary Melvin Laird, "The original Sentinel plan could be misinterpreted as . . . and in fact could have been . . . a first step for the protection of our cities." But in fact there was little technical difference between the Sentinel and Safeguard systems, except that Safeguard would have less effect on suburban real estate values.

The Safeguard system was scotched by doubts about its effectiveness (especially concerning the vulnerability of its radars) and fears about its cost. In 1972 the Nixon administration and the Soviet Union signed the antiballistic missile (ABM) arms control treaty. It limited defenses against ballistic missiles to one hundred interceptors at each of two sites, later reduced by mutual agreement to one hundred interceptors at one site. The site could be located to protect either the national capital or a field of offensive missiles. This would allow the Soviets to maintain their rather primitive Galosh missile defense system around Moscow, while the United States could proceed with the declared aim of the Safeguard system and defend the intercontinental ballistic missile field in North Dakota.

To guard against surprises, the treaty also contained a clause that banned developing, testing, or deploying "ABM systems or components which are sea-based, air-based, space-based, or mobile land-based," a clause that later came under special attack by proponents of missile defense. Despite the proclaimed need for defense of our offensive missiles, neither the Nixon administration nor any following administration maintained the ABM defense of the North Dakota missile field that was allowed under the treaty.

There matters remained until the Reagan administration. It is said that President Reagan was converted to missile defense on a visit to the continental defense headquarters at Cheyenne Mountain, when he was surprised to learn that the United States had no ability to shoot down enemy missiles attacking our country. Be that as it may, in 1983 he announced plans for a Strategic Defense

Initiative, intended to make nuclear weapons "impotent and obsolete." No longer would the system be limited to ground-based interceptor missiles; there were plans for more adventurous technologies, including satellites carrying X-ray lasers that could burn through the skin of an offensive missile booster in the first few minutes after it was launched. The imagined system soon came to be called Star Wars.

Eventually it became clear even to the enthusiasts of the Reagan administration that the X-ray lasers and other features of the Strategic Defense Initiative were beyond current technological capacities. The administration of George Bush Sr. replaced the Strategic Defense Initiative with a system of Global Protection Against Limited Strikes, including about one thousand "brilliant pebbles," small space-based interceptor missiles, along with more conventional land- or sea-based missiles. This strategy also led nowhere, and was allowed to lapse in the Clinton administration.

Research and development continued at a more leisurely pace. In 1996 the Department of Defense announced a plan to continue further development of a scaled-down missile defense system for three years, after which a decision would be made whether or not to deploy the system within the following three years. The National Missile Defense System under study was now limited to a single kind of interceptor missile. Instead of a nuclear weapon, it would carry an "exo-atmospheric kill vehicle" weighing about 120 pounds, which would destroy the enemy warhead above the Earth's atmosphere by a direct hit rather than a nuclear blast. If it worked, it would truly be a bullet hitting a bullet.

Then, on August 31, 1998, North Korea surprised the world by launching a three-stage rocket that carried its third stage over one thousand miles before it broke up into pieces and fell into the Pacific Ocean. The missile did not fly far enough to reach any part of the United States, and it could not have carried a nuclear warhead, but its launch put tremendous political pressure on the Clinton administration to do something soon about missile defense.

In July 1999 President Clinton signed the National Missile Defense Act that had been passed by Congress a few months earlier. Like the Johnson administration's Sentinel initiative, this was more of a defense against Republicans than against external threats. The act committed the United States to deploy a national missile defense "as soon as technologically possible." Later that summer the administration settled on the defense system's initial (C-1) configuration, which remains as a central element of the missile defense system under study by the Bush administration. Twenty (later increased to one hundred) interceptor missiles carrying exo-atmospheric kill vehicles would be based at Fort Greely, Alaska, to be guided to their targets initially by five early-warning radars in Alaska, California, Massachusetts, Greenland, and England. Then, later in their flight, they would be guided by a high-frequency battle management radar on Shemya Island in the Aleutians and finally, in the last six hundred miles of flight, by infrared telescopes carried by the kill vehicles.

This geographical deployment was clearly aimed at defense from North Korean missiles. To better protect the east coast of the United States from missiles launched from the Middle East, it would be necessary later to add interceptor missiles at a second site, perhaps in North Dakota or Maine, and also to add additional battle management radars. The decision to deploy the C-1 system was to have been delayed until 2000, after some of the components of the system had been tested.

The first test of the exo-atmospheric kill vehicle (EKV) was made on October 2, 1999. A dummy warhead that had been sent into space by a Minuteman intercontinental ballistic missile launched from Vandenberg Air Force Base in California was hit over the Pacific by an EKV from an interceptor missile fired from Kwajalein Atoll in the Marshall Islands. But there was less to this success than met the eye. The EKV was at first off course, so that its telescopes did not pick up the Minuteman warhead. When the EKV widened its field of view, at first it saw a large bright balloon decoy and corrected its course, after which it saw the warhead and managed to hit it. If the warhead had not been accompanied by a decoy it might have escaped detection, and if the decoy had looked more like a warhead the EKV would have hit the decoy instead of the warhead. Even so, it seemed that under the right conditions, a bullet could hit a bullet.

Then, in January 2000, the EKV failed a second test. The krypton gas needed to cool the EKV's infrared telescopes had been blocked by ice in the plumbing, so that the EKV never saw the warhead and missed it by over two hundred feet. A third test in

July also failed when the EKV failed to separate from its booster. Further, even if all these tests had been successful, three tests were not nearly adequate to test the system. In August 2000 President Clinton finally decided that the Department of Defense should not start preparing the Alaska site for the battle management radar, and he announced that he would leave the decision whether to deploy the missile defense system to the next administration.

President Bush has taken the movement toward national missile defense in a new direction. Where the Clinton plan called for spending $5.75 billion in 2002 for all forms of ballistic missile defense, the Bush plan calls for spending $8.3 billion on the same tasks. The Bush administration assumed that an antimissile system much like the Clinton administration's National Missile Defense C-1 system would be tested, not only by engagements between rockets fired from Vandenberg and Kwajelein, but also by interceptors fired from Alaska sites, which could later be converted to operational missile defense sites. Also, the Bush administration proposed to supplement this land-based midcourse interception system with an ill-defined mixture of other systems, including possible airborne or spaceborne lasers that would attack enemy missiles during the initial boost phase of their flight.

The Alaska interceptor test site might violate the 1972 ABM treaty; the development and testing of airborne or space-based missile defense systems surely would, long before any actual deployment. But where President Clinton had ruled out a unilateral abrogation of the treaty, President Bush has from the first been eager to free the United States from its restrictions. In August 2001 he said that the United States will withdraw from the treaty at a "time convenient to America." Since then, the disaster of September 11 has brought Presidents Bush and Putin into closer collaboration, but the Russians have refused to agree to major changes in the ABM treaty, and now the president has given notice of his intention to withdraw from it.

2.

So here we are again, arguing the pros and cons of missile defense. The debate raises three main issues.

Would a missile defense system actually protect the United States against

even the sort of attack that might be launched by rogue states like North Ko-
rea or Iraq?

It seems to me likely that the problems that bedeviled the early
tests of the exo-atmospheric kill vehicle can all be solved. The
fourth and fifth tests, in July and December 2001, were successful,
though the kill vehicle booster failed in a test later in December.
The big problem, as it has been since the days of Nike X, is that any
number of interceptor missiles could be used up in attacking decoys that had been sent by the attacker along with its warheads.

This is a particularly acute problem for such missile defense systems as that planned as the first phase of the Clinton-Bush National
Missile Defense, which rely on intercepting warheads in midcourse, above the Earth's atmosphere. Balloons that are deployed
in space at the same time as warheads will follow the same trajec-
tory as the warheads until they reenter the Earth's atmosphere.
They can easily be shaped to look much like warheads to ordinary
telescopes and heated to look like warheads to infrared sensors. It
is also possible and probably even easier to make the warhead look
like a decoy by putting it in a decoy balloon or make it invisible by
hiding it in a cooled shroud. There has been no realistic test of the
ability of an exo-atmospheric kill vehicle to hit a warhead that is accompanied by such penetration aids.

Much of the technical argument over several decades about the
effectiveness of antimissile systems has focused on the question of
whether the United States can solve this problem. It may be that at
great cost we could develop a midcourse interception system that
could defeat any particular group of penetration aids and warheads, but of course we are not likely to know what aids are chosen
by the attacker, and I don't see how we could ever have confidence
in our ability to deal with an unknown threat. The attacker always
has the last move.

Another way of defeating a U.S. midcourse interception system
was mentioned in the report of a "blue ribbon" panel on missile defense convened by Congress in 1998, which was headed by Donald
Rumsfeld and has been emphasized several times since then by
Richard Garwin, one of the panel members. Instead of using a
rocket that would launch at most a few nuclear warheads, an attacker could use the same sort of rocket to launch hundreds of
"bomblets" containing biological warfare agents, such as anthrax

spores. Once deployed, the bomblets would be immune to any sort of missile defense now contemplated. This sort of missile could kill even more people than one carrying a nuclear weapon.

None of these objections applies to a missile defense that can damage an attacker's missile while it is still in "boost phase," i.e., during the brief period when it is being accelerated upward, before it has time to deploy warheads, decoys, or bomblets. But the boost phase lasts only a few minutes. A missile attempting to intercept another missile during the boost phase would have to be launched within about six hundred miles of the intercontinental ballistic missile launch site; so this sort of missile defense system would have to be targeted only at one or at most a few particular potential attackers. For example, a sea-based system that would target missiles in North Korea would not protect against missiles launched from China or Russia. Likewise, an airborne laser would not be effective against missiles that at the end of the boost phase are still beyond the horizon, which for a missile at an altitude of 120 miles is about one thousand miles away. (The actual range of the laser would be substantially less than this.) An air- or sea-based boost phase intercept system could be vulnerable to preemptive attack (as also would the radars of the Clinton-Bush National Missile Defense system), but unless this attack were very carefully timed, it would trigger a counterattack that would destroy the enemy's offensive missiles while they were on the ground. If it could be made to work, a system of space-based lasers (or "brilliant pebbles") might be able to provide protection from threats coming from a much larger area, but this technology does not yet exist, and in any case no specific space-based laser system has been proposed by any administration.

Is it plausible that the United States would be attacked by intercontinental ballistic missiles launched by terrorists or a rogue state, or by accident?

The attack of September 11 made it clear (though it was pretty clear before) that there are people in the world who want to damage us. This seems to have shifted public opinion in favor of missile defense, and it stopped moves in the Senate to deny funding for missile defense tests that would violate the ABM treaty. Hearings on this issue before the Senate Foreign Relations Committee at which I had been asked by the staff to testify were canceled soon after September 11. But the attack also demonstrated that there are ways to hurt the United States that do not involve the launch of intercontinental ballistic missiles. Even nuclear weapons could be delivered in many ways, for instance, by using trucks or freighters or (as suggested by the Rumsfeld panel) ship-launched short-range missiles. But the intercontinental ballistic missile is not just one among the many vehicles that might be used by terrorists or a rogue state to attack us with nuclear weapons — *it is the least likely vehicle.* Though some terrorists are willing to commit suicide in their attacks, the heads of the nations that harbor them never have been. The leaders of the Taliban did not publicly acknowledge that the September 11 attacks were organized in Afghanistan, and Qaddafi has never admitted that the explosion of a Pan American airliner over Lockerbie, Scotland, was planned in Libya.

But unlike such terrorist attacks, an attack by intercontinental ballistic missiles carries an indelible return address. Every launch of such missiles is inevitably detected and its source identified by the fleet of American Defense Support Program satellites. Even granting that a state like North Korea or Iraq might eventually be able to deploy nuclear-armed intercontinental missiles, why would any head of government, however much he may hate us, attack us with intercontinental ballistic missiles, or allow terrorists on his soil to launch such an attack, when he and they could use many other means to deliver nuclear weapons anonymously?

On the other hand, there are circumstances in which the very visibility of intercontinental ballistic missiles might be an advantage. For instance, the United States might not be deterred from its recent actions in Afghanistan or from trying to overthrow Saddam Hussein by a mere threat of nuclear terrorism; but would it risk trying to overthrow a state that had nuclear-armed intercontinental ballistic missiles?

This is a real problem for America, but it is not clear that anything but a perfect antimissile defense would make much difference. If the United States had an antimissile system that had never been used in action, would this give us sufficient confidence to attack a regime that possessed intercontinental ballistic weapons, especially when we did not know what sorts of decoys their missiles carried?

But it need not come to this. There is another way that the United States can avoid being subject to nuclear blackmail by states like Iraq or North Korea. It is occasionally mentioned in discussions of missile defense, though briefly and perhaps with some embarrassment. It is preemption. (Or, as it is sometimes called, pre–boost phase interception.) If a country like Iraq or North Korea were suspected of having nuclear weapons, and we saw that it had tested a ballistic missile of intercontinental range, would we really watch them begin to erect these missiles without taking steps to destroy them on the ground? These steps need not involve our use of nuclear weapons; cruise missiles are now sufficiently accurate to do the job with conventional explosives. I very much doubt if intercontinental ballistic missiles could be put in place by any state without the United States knowing it, and indeed they would be of no use for nuclear blackmail unless they were known to us.

This leaves a mistaken launch by Russia or China as the only plausible way that intercontinental ballistic missiles might threaten the United States. Here "mistake" might mean anything from a purely mechanical malfunction in a single rocket to an unauthorized launch by a few madmen of all of the missiles in a submarine or a land-based missile field, all the way up to the launch of a whole arsenal of missiles ordered by a national leader who is under the mistaken impression that his country is under attack.

Launch by mistake is a serious danger, and although it was not mentioned by President Bush at the time he announced his intention to withdraw from the ABM treaty, it had frequently been cited as one reason for building a national missile defense system. Indeed, a large-scale mistaken attack by Russia is the only plausible threat that could not only damage our country but destroy it beyond our ability to recover. Such an attack would be far more devastating than anything terrorists could manage. Russia has some thirty-nine hundred strategic nuclear warheads, of which over one thousand are on land- or submarine-based intercontinental ballistic missiles that are ready to be launched at a moment's notice. These missiles are increasingly vulnerable to an American first strike, as the Russian early warning capabilities get progressively weaker. At least twice the Russians have mistakenly thought that they were under missile attack: once in March 1983 because a Soviet satellite mistook bright reflections for the launch of five missiles, and again in January 1995 because a Norwegian research rocket that had been detected by a Russian radar was interpreted as an incoming American missile. In both cases, the Russian launch process came within minutes of the point where they would make a decision whether or not to launch a nuclear retaliation.

The danger of a launch during a crisis is made worse by the fact that most Russian and American warheads are MIRVs — multiple independently targeted reentry vehicles. For example, each Russian SS-18 missile carries ten warheads, each of which can be directed to a separate target. Without MIRVs, and with equal numbers of missiles on each side, there would be a disadvantage in striking first. Even if every missile had a 90 percent chance of destroying its target, the side that struck first even with all its missile forces targeted at its adversary's missiles would leave its own arsenal empty while its adversary would still have 10 percent of its forces left. But with, say, ten warheads on each missile, the side that struck first with just 10 percent of its forces could destroy 90 percent of the adversary's forces and still have 90 percent of its own forces left. Of course, this reasoning is insane. No one today thinks that either Russia or the United States would plan such an attack. But in some future crisis, with different Russian leaders and with misleading data coming in from early warning radars — who knows?

The sort of missile defense planned by the Bush administration would not protect us against a massive attack by mistake. Indeed, it has been specifically advertised not to be able to defend us from a large-scale Russian attack. (It might not even protect us against a mistaken launch of a few missiles, since Russian missiles are presumably accompanied by sophisticated decoys or other penetration aids; they surely would be if we were to deploy a missile defense system.) With or without the Bush missile defense plan, we have to face the danger of annihilation by Russian nuclear-armed missiles. With the degradation of Russian early warning capacities and the general loosening of Russian society, this danger may be even greater than it was during the cold war. Which brings me to the third and most important issue.

Would a national missile defense system of the sort proposed by the Bush administration help or hurt our national security?

At first sight, this question seems to answer itself. Isn't any missile defense, however ineffective, better than no missile defense at all?

One trouble with this reasoning is that we do not face a fixed threat, a threat independent of what we do about missile defense.

True, we are not now in the position we were in during the 1960s and 1970s, when we could reasonably expect that any U.S. missile defense system would be countered by an increase in Soviet offensive missile forces. The current Russian economy would not support an increase in Russia's missile forces; indeed, the Russians have been eager to reduce their forces, reportedly down to some two thousand or so strategic nuclear warheads. But this is still a force that could destroy the United States, and much else in the world besides. The large size of their arsenal also increases the danger that Russian nuclear weapons or even long-range missiles might be stolen or sold to terrorists or rogue states. I am told that Russia now maintains tight control over its strategic nuclear weapons, but this wasn't true in the early 1990s, and it may not be true in future. There is nothing more important to American security than to get nuclear forces on both sides down at least to hundreds or even dozens rather than thousands of warheads, and especially to get rid of MIRVs, but this is not going to happen if the United States is committed to a national missile defense.

The Russian nuclear force is the sole remnant of its status as a superpower. Whatever good feelings may exist now between us and Russia, any U.S. system that might defend our country against even a few Russian intercontinental ballistic missiles therefore sets a limit below which the Russians will not go in reducing their strategic nuclear forces. Even if Russia is forced by economic pressures to continue reducing its missile forces, it can cheaply maintain its deterrent, although in ways that are dangerous. It could, for example, remove whatever inhibitions it may now have from launching its missiles on a moment's notice. Nor is Russia likely to eliminate its MIRVs if the United States goes ahead with missile defense. The START II treaty was to have eliminated all land-based MIRVs on both sides, but the Russians have already indicated that they will not go through with implementing this treaty if the United States abrogates the ABM treaty.

As for China, it has right now about twenty nuclear-armed intercontinental ballistic missiles, enough for a significant deterrent against any attack from Russia or the United States. A recent National Intelligence Estimate that was leaked to the *New York Times* and the *Washington Post* predicted that if the United States develops a national missile defense, then the Chinese will increase their forces from about twenty to about two hundred missiles. And if China makes this sort of increase in its missile force, then what will Japan and India do? And then what will Pakistan do?

It may seem contradictory to argue that the proposed national missile defense system would probably be ineffective against even a small attack by a rogue state, while also arguing that it would prevent needed reductions in Russian missile forces and promote increases in Chinese missile forces. But each country "prudently" tends to overestimate the effectiveness of any other country's defenses, especially as they may develop in future. The Soviet deployment of a primitive antimissile defense of Moscow was a major factor in America's decision to multiply its warheads by deploying MIRVs, so that Moscow was in more danger after it was defended than before. I remember how in the early 1970s U.S. defense planners became terrified that Soviet antiaircraft missiles might be given a role in defense against intercontinental ballistic missiles, something that never happened. Imagine then how Russians and the Chinese defense planners will take account of the unilateral American withdrawal from the 1972 ABM treaty.

If it were possible tomorrow to switch on a missile defense system that would make the United States invulnerable to any missile attack, then I and most other opponents of missile defense would be all for it. But that is not the choice we face. What is at issue is a missile defense system that will take almost a decade to deploy in its initial phase, and then many more years to upgrade to the point where at best it would have some effectiveness against some plausible threats. During all of this time, however, American security will be damaged by measures taken by Russia or China to preserve or enlarge their strategic capability in response to our missile defense.

There is one sort of missile defense that would not raise these problems. As already mentioned, a defense that targets intercontinental ballistic missiles during the boost phase with either missiles or airborne laser beams could only defend against missile launches within a limited geographical area; and it would also be immune to decoys and other penetration aids. We could defend against the launch of North Korean missiles by using short-range missiles based on ships in the Sea of Japan, though to defend against a launch from Iraq or Iran would require cooperation from Turkey or some republic of the former Soviet Union, respectively. Such a defense would have no effectiveness against missiles launched from sites in Russia or China, which during the boost phase would be beyond the range of any missiles or airborne lasers we might deploy. For this reason, although this sort of defense would violate the 1972 ABM treaty, President Putin has already indicated that he would consider revising the treaty to allow it. But a boost phase intercept system would have all the destabilizing effects of other missile defense systems if it were based on satellites or if it were combined with exo-atmospheric midcourse interceptors like those of the Clinton-Bush National Missile Defense proposal.

Developing a national missile defense system would also harm our foreign relations. It would add to the general perception that the United States is unwilling to be bound by international agreements, such as comprehensive test ban treaties or environmental agreements. It would weaken Putin's hand in dealing with Russian ultranationalists. By trying to defend the United States from missile attacks while leaving our allies defenseless, missile defense would tend to undermine alliances like NATO. A boost-phase intercept system would not really be an exception; it is true that the interception of a long-range missile in the boost phase does not depend much on the destination of the missile, but by interrupting the boost it would probably only cause the warhead to fall short, perhaps on an ally such as Canada or Germany.

A missile defense system would hurt our security in another important way, by taking money away from other forms of defense. We are simply unable to do everything we can imagine that might defend us. We need to upgrade our hospitals to deal with biological attack; improve security along our border with Canada and in our ports; upgrade the FBI computer system; and so on. All of our activities along these lines are constrained by a lack of funds. Legislation to increase funding for homeland defense was blocked in the House of Representatives because the amounts of money requested exceeded the administration's guidelines.

If we are particularly worried (as we should be worried) about terrorist nuclear attacks on the United States, then we ought to give a very high priority to working with Russia and other countries to get rid of the large stocks of weapons-grade plutonium and uranium that are produced by their power reactors. Russia now holds about 150 tons of plutonium and 1,000 tons of highly enriched uranium. This material could be used not only to make nuclear bombs, which can be delivered to the United States in all sorts of ways; even a technically unsophisticated terrorist could instead use it to make so-called dirty bombs, in which an ordinary high explosive is surrounded with highly radioactive material that when dispersed in an explosion would make large urban areas uninhabitable.

Unfortunately, this material is not under tight control. Since 1991 there has been a bipartisan Nunn-Lugar Cooperative Threat Reduction Program that, among other things, aims at improving the security of Russian control over fissionable materials and making Russian plutonium and uranium unusable as nuclear explosives, but this too is not being adequately funded. A bipartisan panel headed by Howard Baker and Lloyd Cutler has called for spending at an average level over the next decade of about $3 billion a year for securing, monitoring, and reducing Russian nuclear weapons, materials, and expertise. The amount in the 2002 budget for these activities is only about $750 million, even after substantial increases by Congress. Comparison of these figures with the $60 billion quoted cost (certain to be greatly exceeded) of a minimum missile defense system gives a powerful impression that the Bush administration and some in Congress are not entirely serious about national security.

I was at a press conference in Washington in November 2001, when the Federation of American Scientists released a letter signed by fifty-one Nobel laureates that opposed spending on national missile defense programs that would violate the 1972 ABM treaty. One of the reporters present asked me why, if the arguments against national missile defense are so cogent, many people in and out of government are for it. It was a good question, and one to which I am not sure I know the answer. There are the usual pressures for large military programs that come from defense contractors and from politicians trading on patriotism. The arguments for national missile defense may seem simpler and more straightforward than the arguments against it. But I think there is also a peculiar fascination with anything that projects American power into space. How else explain the idiocy of the International Space Station, or the card tables that were set up at airports during the Reagan administration by people advocating a "high frontier" missile defense program? I have to admit that thoughtlessness is not a monopoly of missile defense advocates. Some opponents of missile defense are automatically against any large military program. In assessing missile defense, or anything else for that matter, there is no substitute for actually thinking through the issues.

In my own field of physics, we make a distinction between applied physics, which is motivated by some social need, and pure physics, the search for knowledge for its own sake. Both kinds of physics are valuable, but not everything pure is desirable. In seeking to deploy a national missile defense aimed at an implausible threat, a defense that would have dubious effectiveness against even that threat, and that on balance would harm our security more than it helps it, the Bush administration seems to be pursuing a pure rather than applied missile defense — a missile defense that is undertaken for its own sake rather than for any application it may have in defending our country.

TED WILLIAMS

Maine's War on Coyotes

FROM *Audubon*

IN MY MEADOW — in back of the computer screen, amid lupine, phlox, high grass, and bluebird boxes — things happen that make writing difficult. On a summer morning a wild canid, sleek coat gleaming in the sun, leaps high, twists, swats the ground with huge forepaws. He is well muscled but, in late puppyhood, still gawky and playing as much as seriously hunting mice. His race, maybe not new but newly noticed, first got the attention of humans about sixty years ago. It has been called "coy dog," "coy wolf," "new wolf," "brush wolf," and, now, "eastern coyote."

The "coyote" part is misleading. Eastern coyotes are larger, heavier-set, and much more wolflike than their western cousins. They possess wolf genes, so maybe they interbred with wolves on their way east. Or maybe they were here all along, identified as small wolves whenever they were shot, trapped, or poisoned by the many people who hated them. When moonlight washes hardwood ridges, I like to howl at eastern coyotes; they answer me, then embark on prolonged conversations among themselves. Unfortunately for midlevel plant communities, shrub-nesting birds, and my wife's tulips, they kill very few deer in central New England.

They kill more deer in Maine, but not enough to limit the population, which has been growing for twenty years. Maine is the only state that sees fit to hire eastern coyote–control agents for the alleged benefit of deer hunters and one of the few states south of Alaska that still believes it's desirable or even possible to make more game by knocking off predators. Unlike wolves, eastern coyotes prefer low-energy pursuit. They'll take deer when varying hare (their favorite prey) are scarce or when they can do so safely and economically, as when the snow is deep and crusted and the deer are "yarded up" in thick conifers. But in the colder regions, where deer haven't been increasing, the limiting factor is poor winter habitat, not coyotes.

In June 1980, on my first assignment for *Audubon,* I joined a team from Maine's Cooperative Wildlife Research Unit and followed radio-collared eastern coyotes around Maine's western mountains. As far as the unit had been able to determine, in nine months forty-one animals had killed zero deer. There had been deer hair in the scats, but in almost every case the researchers had been able to pinpoint the source: carrion. Documented stomach or scat contents had included varying hare, butterflies, beetles, grasshoppers, berries, apples, offal, fish, voles, aluminum foil, rope, leaves, leather, dog food, and dog food bags. That year eastern coyotes made it onto the Maine Republican platform, where they were identified as one of just three environmental ills worthy of the party's consideration (the others being anti-pesticide sentiment and unfair property taxes).

At least with coyotes there was bipartisan agreement. When I had finished my work in the woods, I stopped by Augusta to visit Glenn Manuel, the father of eastern coyote control. For helping get the state's Democratic governor elected, Manuel, a former state senator, had been appointed commissioner of the Maine Department of Inland Fisheries and Wildlife. He told me that coyotes threatened to eliminate Maine's deer herd and scoffed at his biologists who claimed otherwise.

"[They] still believe in the balance of nature," he declared. "They're textbook boys." Later, without a shred of documentation, he publicly announced that "many does are found dead, but only the unborn fawn is eaten" and lamented that coyote snares were illegal. Even as commissioner, Manuel ardently supported the Dickey-Lincoln dams project, which would have destroyed 130,000 acres of habitat for deer and other wildlife, 268 miles of free-flowing trout water, and thirty wilderness ponds. And when he chaired a public meeting to consider reintroducing wildlife that had been extirpated from Maine, his biologists hid behind their clipboards when he urged them to "bring back" penguins.

*

Manuel, to borrow lyrics from the folksinger Tom Rush, would have killed "a thousand coyotes, if [he] could only just find one." In 1980, the first full year of the control program he designed, the entire Maine warden force "controlled" three animals. But five years later the state legislature ordered the department to kill coyotes, authorizing it to hire private citizens as coyote control agents and train them in the use of snares, now legal. About 60 of these agents have been snaring roughly 400 coyotes per season (winter and early spring). Last season, 51 killed 564, presumably as a result of liberalized regulations implemented after the legislature passed a resolution asking the department "to encourage the harvest of coyotes."

The public is unhappy about this. In Augusta, at the June 1, 2002, meeting of a group of angry citizens who call themselves the Nosnare Task Force, Susan Cockrell showed me a coyote snare. Basically, it's a noose made of stout cable. You hang it from a tree, and when the coyote sticks its head through the loop, it closes on the animal's neck. A floppy washer keeps the loop from loosening. Cockrell, one of the group's founders, teaches nature writing at the University of Maine at Orono. Other founders include her husband, Will La Page, a forestry professor at the same institution; the wildlife biologist Debra Davidson; and a registered Maine guide, Daryle DeJoy.

They plied me with internal correspondence they had excavated from the bowels of the department under threat of Maine's Sunshine Law, a state version of the Freedom of Information Act. As I perused the reports, memos, and e-mails, the value of the law became increasingly apparent. Without it, the public would never know what wildlife biologists think about coyote control or how the state legislature had secretly and successfully pressured the department into liberalizing the regs. Lawmakers had opted for a resolution rather than a bill because resolutions don't require public participation.

Department biologists repeatedly observe that killing coyotes stimulates reproduction and that in order to lower a population, you have to remove at least 70 percent of the animals every year. But even with the relaxed regs, the control agents are getting only 4 percent. In areas off-limits to coyote controllers (Yellowstone National Park and the Hanford nuclear site in Washington State, for example), an average of fewer than two pups make it to fall. In "normal areas," where humans are busy killing the coyotes' main competition — i.e., other coyotes — the figure is about six. Moreover, coyotes are highly territorial. An alpha male might be defending a cedar swamp, killing deer when conditions are right — say, 4 a winter. If that coyote is snared, half a dozen subordinate coyotes might move in and kill 24 deer. In 1946 federal agents killed 294,000 coyotes in seventeen western states. In 1974 — after twenty-eight years of intensive trapping, shooting, and poisoning — they killed 295,000 in the same seventeen states. When populations increase, so do ranges. Some scientists believe that coyote persecution in the West is why there are coyotes in the East.

"Coyote snaring is a mean-spirited government program whose sole intent is to catch and strangle wildlife with a wire noose, for some perceived biological gain," Chuck Hulsey, one of Maine's seven regional wildlife biologists, told me, emphasizing that he was speaking for himself and not his department. "You cannot stockpile deer like money in a mutual fund, to be enjoyed at a later date. Spending many tens of thousands of dollars to snare a few hundred coyotes . . . is a poor use of public dollars."

Among wildlifers it is considered "unprofessional" to fret about humane issues. But there's a limit; when cruelty to wild animals becomes sufficiently severe and senseless, good biologists get involved. "Killing an animal by strangling it with a wire loop often results in a slow, painful death, sometimes lasting days . . ." wrote Hulsey to his bureau director. "It would violate state humane laws to treat a domestic dog in the same manner."

Hulsey is just one of many department biologists speaking out. Last fall Wally Jakubas, the agency's top mammal scientist, got concerned when, checking 94 snared coyotes during a study to determine the genetics of the beast, he noticed a large proportion of carcasses with grotesquely swollen heads, bullet holes, fractured limbs, and broken teeth. Of particular interest to Jakubas were the animals with swollen heads — "jellyheads," the snarers call them. When the snare doesn't close sufficiently, it constricts the jugular vein on the outside of the neck, cutting off blood returning to the heart; meanwhile, the carotid artery keeps pumping blood into the brain, eventually rupturing its vascular system. In a memo to his supervisor, Jakubas wrote: "I think it is also safe to say that [this] is an unpleasant death. Anyone who has had a migraine knows what it feels like to have swollen blood vessels in the head. To have blood vessels burst because of pressure must be excruciating." Almost a third of the animals Jakubas looked at were jellyheads. Almost another third had been clubbed or shot, indicating that, contrary to department claims, the snares hadn't killed them quickly. Coyote control agents have to check their snares only every three days, and under the liberalized regs suggested by the legislature, they can get permission to check them only every seven days.

Jakubas promptly turned his report over to his superiors, who promptly sat on it. Eventually someone leaked it to Maine Public Radio, thereby setting the Sportsman's Alliance of Maine and the Maine Trappers Association into full cry. These outfits, which together make up the state's *entire* coyote control lobby and which claim (falsely) to speak for Maine's hunters, anglers, and trappers, crammed their publications and Web sites with screeds about the alleged treachery of Jakubas, the alleged incompetence of his help, and the alleged deficiencies of his study. He had revealed facts they didn't want to know and, especially, didn't want the public to know. Until this information got daylighted, the only thing they had to do to perpetuate the boondoggle was hiss into the ears of the thirteen lawmakers who sit on the Joint Standing Committee on Fisheries and Wildlife.

Howard Chick of Lebanon, Maine, a member of the Sportsman's Alliance, has hunted deer since the 1930s. He was born in 1922, in the farmhouse where he lives. In 1881 his father was born in this farmhouse. In 1843 his grandfather was born in this farmhouse. The farmhouse is on Chick Road. Howard Chick "dispatches coyotes when they show themselves." But somehow there are never any fewer. He doesn't buy the balance-of-nature stuff. "These are things I don't have to have a biologist tell me," he proclaims, in reference to the department's assertion that you have to annually remove 70 percent of a coyote population to reduce it. "Suppose you had a dozen rattlesnakes in your immediate vicinity," he says. "Any one you dispatch is going to lessen the chances of your getting bit; it's the same with coyotes." But it isn't. Like so many other Maine deer hunters, Howard Chick doesn't understand that killing coyotes is like trying to put out a fire with kerosene. You can do it if the fire is small and you have lots of kerosene, but coyote populations are never small and, in comparison, the amount of control is always tiny.

In southern and central Maine there are now so many deer that in some areas they're damaging their range, but Howard Chick worries more about deer in the north, where they are less plentiful. This is the natural range of moose and caribou, and deer are here mostly because humans have created openings in the boreal forest. Even with thick conifers to provide thermal cover, deer get winter-stressed in these parts. And now a lot of that cover has been re-moved by spruce budworm and paper companies. The official line from Fisheries and Wildlife is that coyote control in the north woods, in specific deeryards, "may" result in temporary relief for wintering deer. But it also may not. The department doesn't know, because it hasn't done any research. "In northern, western, and eastern sections of Maine, inadequate wintering habitat is the pri-mary factor limiting deer populations," writes Maine's deer biolo-gist, Gerald Lavigne. "There, high predation rates by coyotes are the symptoms, not the cause, of deer population problems."

Lavigne had been responding to the state legislature, which in 1995 ordered the department to "conduct a study to determine the impact that coyotes have on deer, and to propose recommen-dations to encourage the harvest of coyotes." The bill had been sponsored by Howard Chick, who, in addition to being a farmer and a deer hunter, is a state representative, a member of the Joint Standing Committee on Fisheries and Wildlife, and the oldest member of the Maine legislature. Howard Chick, in fact, is the rea-son Maine now has paid recreational coyote snaring. Not believing the stuff he read in Lavigne's report, he introduced the 2001 reso-lution that hatched the liberalized snaring regulations. Before the resolution, coyote snaring (however misguided) had been in re-sponse to observed deer mortality. Now it's in response to the whims of the snarer.

It is curious behavior for the public to pay for the training of pro-fessional wildlife managers at state universities, pay their salaries, pay their expenses, and then pay politicians to tell them how to manage wildlife. According to Maine law, the Department of In-land Fisheries and Wildlife "shall maintain a coyote control program." It has no choice, but that excuse carries it only so far. The law also provides leeway: "The commissioner may employ qualified persons to serve as agents of the department for the purposes of coyote control." There's nothing in there that says he has to. Glenn Manuel, who thought penguins belong in Maine, was a career potato farmer. Lee Perry — the current commissioner, appointed in the fall of 1997 — is a career wildlife biologist. Wildlife advocates expect more from him, especially now that the Nosnare Task Force has shown them the nasty realities of coyote control. As a first step, Perry could order his information and education staff to drop everything and start disabusing Maine deer hunters of their copious superstitions. But instead of leading and educating, the department plays subordinate coyote to the Sportsman's Alliance of Maine and the Maine Trappers Association, rolling over and peeing on itself whenever it gets barked at.

In response to Chick's resolution, the department organized an ad hoc "study group" to make recommendations for new snaring regs. But of the groups that participated — the Sportsman's Alliance of Maine, the Maine Trappers Association, coyote control agents, the department, and the Maine Audubon Society — only the last disapproved of snaring. "The so-called study consisted of one meeting and one phone call," complains a Maine Audubon biologist, Jody Jones. "The department took none of our advice. One thing that really upset me was that in the commissioner's form letter responding to letters and e-mails critical of the snaring program, he said Maine Audubon had participated in this group and these were the recommendations that came forward. That wasn't a lie, but the implication was that we supported the snaring program. We got angry calls from members."

The department also ignored a lot of advice from its own biologists, who had expressed concern for the nontarget wildlife that have been found dead in coyote snares — eagles, deer, moose, bears, fishers, foxes, bobcats, and especially Canada lynx, now federally threatened. They had asked that snaring not be conducted in March, when so much of this wildlife is on the move. They had objected to the proposed regulation that did away with the limit on the number of snares an agent can set. (Since snares cost less than a dollar each, there's scant motivation to collect them when the season is over.) They had asked that snaring not be done where lynx had been seen and in lynx study areas. In every one of these cases they were overruled.

On the other hand, the department acquiesced to George Smith, director of the Sportsman's Alliance of Maine, who had written to Perry as follows: "The [snaring] limits are apparently proposed to appease federal officials and radical environmental groups now that the Canadian [*sic*] lynx has been listed . . . We cannot support a policy that puts lynx ahead of deer in the north woods . . . Our suspicions that DIF&W is not really committed to this program are only enhanced by this most recent decision [now revoked] to stop protecting deer in one of the largest and most important deeryards in the entire north woods."

The department's report to the legislature in response to the resolution listed twenty-five concerns of the ad hoc study group, followed by detailed explanations from the department. Concern No. 11 was "lack of support for snaring throughout MDIFW." But in the draft report, the department declined to provide a single reason for this lack. When Hulsey suggested to his bureau director that the public deserved an explanation, he was told to write one. He complied, with a five-page memo that spared no detail. But the final report contained not a word of that explanation or of any other.

Whenever the public expresses concern about the threat to lynx, the department responds that no snarer has reported killing a lynx since the species was listed. *Of course* no snarer has reported killing a lynx. Such a confession could elicit prosecution under the Endangered Species Act. "We've had open discussions about the dangers of coyote snaring in lynx areas," says Paul Nickerson, chief of threatened and endangered species for the U.S. Fish and Wildlife Service's northeast region. "So if a lynx is taken, no one can just say, 'Oops.'"

The official word from the department — repeatedly contradicted by internal correspondence from its own biologists — is that the coyote-snaring program is legitimate animal damage control, *not* paid recreation. But animal damage control is, as the name implies, a response to animal damage. You don't do it in advance. The department doesn't go around knocking off bears because they might one day tip over a beehive; it doesn't eradicate beavers in spring because they might flood someone's cellar the following winter.

However, the days when Maine wildlife could be managed by politicians like Howard Chick and radical special interest groups like the Sportsman's Alliance of Maine and the Maine Trappers Association appear to be drawing to a close. For one thing, the Nosnare Task Force has alerted the public to the cruelty and stupidity of coyote control. Of recent comments received by the department from Maine residents, 7 supported coyote snaring, 4 were undecided, and 77 were opposed. Because the department is funded almost exclusively by hunters, fishermen, and trappers, it hasn't had to pay a lot of attention to anyone else. But that's changing too. Next year 18 percent of its budget will come from the general fund, and with public funding comes public representation.

It is not clear how much of a threat eastern coyote "control" is to nontarget species. It is abundantly clear that it is no threat whatsoever to the eastern coyote. Maybe what it threatens most is the reputation of legitimate, ethical sportsmen who already are getting kicked around by the animal rights crowd. As Mark McCullough, the Department of Inland Fisheries and Wildlife's endangered species leader, has advised his supervisor: "All it will take will be one animal rights advocate to videotape a 'jellyhead' in a snare and your program will be over, and maybe even take recreational trapping with it."

Moving into Maine with the coyotes themselves is the image of the Western coyote controller, eloquently captured by Tom Rush. You wouldn't have heard "A Cowboy's Paean to a Coyote" (a paean, pronounced "pee-in," is a song of joyful praise or exultation). Rush wrote it when he lived in Wyoming, to commemorate a three-day coyote shoot in which several hundred participants came up with a total of two coyotes, one with tire tracks on it. Herewith, a few verses:

> Go on out and shoot yourself some coyotes,
> Makes a man feel good, Lord, it makes a man feel proud!
> Go on out and shoot yourself some coyotes,
> One for Mother, one for Country, one for God.
> Well, if you're having trouble with the truck, or with the woman,
> Maybe them kids are screwin' up in school.
> If the cows are actin' smarter than the cowboy,
> You gotta show the world you ain't nobody's fool.
> I got my field rations straight from old Jack Daniel's,

Hank, Jr.'s on the eight-track in my four-by-four.
And I'd shoot a thousand coyotes if I could only just find one,
'Cause, boys, that's what God made coyotes for.

Maine coyote control wastes something much more valuable than time, money, or even the sportsman's image. It wastes the credibility, effectiveness, and morale of an otherwise enlightened agency that is doing superb work restoring native ecosystems.

This year the Maine Department of Inland Fisheries and Wildlife has an $8 million deficit, and Governor Angus King, who appears oblivious to the bad image coyote control is giving his state, has asked it to come up with ways of cutting back on expenditures. Under the liberalized snaring regulations, the costs of administering the coyote control program have about tripled, at least according to one internal estimate. The governor needs to pay more attention to the people who truly know, the people who make the recommendations that get ignored by the decision makers, the people the public doesn't hear from except when someone rifles through dusty file cabinets — Maine's wildlife biologists. For the first budget cut, every one of them would have the same recommendation.

EDWARD O. WILSON

The Bottleneck

FROM *Scientific American*

THE TWENTIETH CENTURY was a time of exponential scientific and technical advance, the freeing of the arts by an exuberant modernism, and the spread of democracy and human rights throughout the world. It was also a dark and savage age of world wars, genocide, and totalitarian ideologies that came dangerously close to global domination. While preoccupied with all this tumult, humanity managed collaterally to decimate the natural environment and draw down the nonrenewable resources of the planet with cheerful abandon. We thereby accelerated the erasure of entire ecosystems and the extinction of thousands of million-year-old species. If Earth's ability to support our growth is finite — and it is — we were mostly too busy to notice.

As a new century begins, we have begun to awaken from this delirium. Now, increasingly post-ideological in temper, we may be ready to settle down before we wreck the planet. It is time to sort out Earth and calculate what it will take to provide a satisfying and sustainable life for everyone into the indefinite future. The question of the century is: How best can we shift to a culture of permanence, both for ourselves and for the biosphere that sustains us?

The bottom line is different from that generally assumed by our leading economists and public philosophers. They have mostly ignored the numbers that count. Consider that with the global population past 6 billion and on its way to 8 billion or more by mid-century, per capita freshwater and arable land are descending to levels resource experts agree are risky. The ecological footprint — the average amount of productive land and shallow sea appropriated by each person in bits and pieces from around the world for food, water, housing, energy, transportation, commerce, and waste absorption — is about 1 hectare (2.5 acres) in developing nations but about 9.6 hectares (24 acres) in the United States. The footprint for the total human population is 2.1 hectares (5.2 acres). For every person in the world to reach present U.S. levels of consumption with existing technology would require four more planet Earths. The 5 billion people of the developing countries may never wish to attain this level of profligacy. But in trying to achieve at least a decent standard of living, they have joined the industrial world in erasing the last of the natural environments. At the same time, *Homo sapiens* has become a geophysical force, the first species in the history of the planet to attain that dubious distinction. We have driven atmospheric carbon dioxide to the highest levels in at least 200,000 years, unbalanced the nitrogen cycle, and contributed to a global warming that will ultimately be bad news everywhere.

In short, we have entered the Century of the Environment, in which the immediate future is usefully conceived as a bottleneck. Science and technology, combined with a lack of self-understanding and a Paleolithic obstinacy, brought us to where we are today. Now science and technology, combined with foresight and moral courage, must see us through the bottleneck and out.

"Wait! Hold on there just one minute!"

That is the voice of the cornucopian economist. Let us listen to him carefully. He is focused on production and consumption. These are what the world wants and needs, he says. He is right, of course. Every species lives on production and consumption. The tree finds and consumes nutrients and sunlight; the leopard finds and consumes the deer. And the farmer clears both away to find space and raise corn — for consumption. The economist's thinking is based on precise models of rational choice and near-horizon timelines. His parameters are the gross domestic product, trade balance, and competitive index. He sits on corporate boards, travels to Washington, occasionally appears on television talk shows. The planet, he insists, is perpetually fruitful and still underutilized.

The ecologist has a different worldview. He is focused on unsustainable crop yields, overdrawn aquifers, and threatened ecosystems. His voice is also heard, albeit faintly, in high government and corporate circles. He sits on nonprofit foundation boards, writes for *Scientific American,* and is sometimes called to Washington. The planet, he insists, is exhausted and in trouble.

The Economist

"Ease up. In spite of two centuries of doomsaying, humanity is enjoying unprecedented prosperity. There are environmental problems, certainly, but they can be solved. Think of them as the detritus of progress, to be cleared away. The global economic picture is favorable. The gross national products of the industrial countries continue to rise. Despite their recessions, the Asian tigers are catching up with North America and Europe. Around the world, manufacture and the service economy are growing geometrically. Since 1950 per capita income and meat production have risen continuously. Even though the world population has increased at an explosive 1.8 percent each year during the same period, cereal production, the source of more than half the food calories of the poorer nations and the traditional proxy of worldwide crop yield, has more than kept pace, rising from 275 kilograms per head in the early 1950s to 370 kilograms by the 1980s. The forests of the developed countries are now regenerating as fast as they are being cleared, or nearly so. And while fibers are also declining steeply in most of the rest of the world — a serious problem, I grant — no global scarcities are expected in the foreseeable future. Agriforestry has been summoned to the rescue: more than 20 percent of industrial wood fiber now comes from tree plantations.

"Social progress is running parallel to economic growth. Literacy rates are climbing, and with them the liberation and empowerment of women. Democracy, the gold standard of governance, is spreading country by country. The communication revolution powered by the computer and the Internet has accelerated the globalization of trade and the evolution of a more irenic international culture.

"For two centuries the specter of Malthus troubled the dreams of futurists. By rising exponentially, the doomsayers claimed, population must outstrip the limited resources of the world and bring about famine, chaos, and war. On occasion this scenario did unfold locally. But that has been more the result of political mismanagement than Malthusian mathematics. Human ingenuity has always found a way to accommodate rising populations and allow most to prosper.

"Genius and effort have transformed the environment to the benefit of human life. We have turned a wild and inhospitable world into a garden. Human dominance is Earth's destiny. The harmful perturbations we have caused can be moderated and reversed as we go along."

The Environmentalist

"Yes, it's true that the human condition has improved dramatically in many ways. But you've painted only half the picture, and with all due respect, the logic it uses is just plain dangerous. As your worldview implies, humanity has learned how to create an economy-driven paradise. Yes again — but only on an infinitely large and malleable planet. It should be obvious to you that Earth is finite and its environment increasingly brittle. No one should look to gross national products and corporate annual reports for a competent projection of the world's long-term economic future. To the information there, if we are to understand the real world, must be added the research reports of natural-resource specialists and ecological economists. They are the experts who seek an accurate balance sheet, one that includes a full accounting of the costs to the planet incurred by economic growth.

"This new breed of analysts argues that we can no longer afford to ignore the dependency of the economy and social progress on the environmental resource base. It is the content of economic growth, with natural resources factored in, that counts in the long term, not just the yield in products and currency. A country that levels its forests, drains its aquifers, and washes its topsoil downriver without measuring the cost is a country traveling blind.

"Suppose that the conventionally measured global economic output, now at about $31 trillion, were to expand at a healthy 3 percent annually. By 2050 it would in theory reach $138 trillion. With only a small leveling adjustment of this income, the entire world population would be prosperous by current standards. Utopia at last, it would seem! What is the flaw in the argument? It is the environment crumbling beneath us. If natural resources, particularly fresh water and arable land, continue to diminish at their present per capita rate, the economic boom will lose steam, in the course of which — and this worries me even if it doesn't worry you — the effort to enlarge productive land will wipe out a large part of the world's fauna and flora.

"The appropriation of productive land — the ecological footprint — is already too large for the planet to sustain, and it's growing larger. A recent study building on this concept estimated that the human population exceeded Earth's sustainable capacity around the year 1978. By 2000 it had overshot by 1.4 times that capacity. If 12 percent of land were now to be set aside in order to protect the natural environment, as recommended in the 1987 Brundtland Report, Earth's sustainable capacity will have been exceeded still earlier, around 1972. In short, Earth has lost its ability to regenerate — unless global consumption is reduced or global production is increased, or both."

By dramatizing these two polar views of the economic future, I don't wish to imply the existence of two cultures with distinct ethos. All who care about both the economy and environment, and that includes the vast majority, are members of the same culture. The gaze of our two debaters is fixed on different points in the space-time scale in which we all dwell. They differ in the factors they take into account in forecasting the state of the world, how far they look into the future, and how much they care about nonhuman life. Most economists today, and all but the most politically conservative of their public interpreters, recognize very well that the world has limits and that the human population cannot afford to grow much larger. They know that humanity is destroying biodiversity. They just don't like to spend a lot of time thinking about it.

The environmentalist view is fortunately spreading. Perhaps the time has come to cease calling it the "environmentalist" view, as though it were a lobbying effort outside the mainstream of human activity, and to start calling it the real-world view. In a realistically reported and managed economy, balanced accounting will be routine. The conventional gross national product (GNP) will be replaced by the more comprehensive genuine progress indicator (GPI), which includes estimates of environmental costs of economic activity. Already a growing number of economists, scientists, political leaders, and others have endorsed precisely this change.

What, then, are essential facts about population and environment? From existing databases we can answer that question and visualize more clearly the bottleneck through which humanity and the rest of life are now passing.

On or about October 12, 1999, the world population reached 6 billion. It has continued to climb at an annual rate of 1.4 percent, adding 200,000 people each day or the equivalent of the population of a large city each week. The rate, though beginning to slow, is still basically exponential: the more people, the faster the growth, thence still more people sooner and an even faster growth, and so on upward toward astronomical numbers unless the trend is reversed and the growth rate is reduced to zero or less. This exponentiation means that people born in 1950 were the first to see the human population double in their lifetime, from 2.5 billion to over 6 billion now. During the twentieth century more people were added to the world than in all of previous human history. In 1800 there had been about 1 billion and in 1900, still only 1.6 billion.

The pattern of human population growth in the twentieth century was more bacterial than primate. When *Homo sapiens* passed the 6 billion mark, we had already exceeded by perhaps as much as 100 times the biomass of any large animal species that ever existed on the land. We and the rest of life cannot afford another hundred years like that.

By the end of the century some relief was in sight. In most parts of the world — North and South America, Europe, Australia, and most of Asia — people had begun gingerly to tap the brake pedal. The worldwide average number of children per woman fell from 4.3 in 1960 to 2.6 in 2000. The number required to attain zero population growth — that is, the number that balances the birth and death rates and holds the standing population size constant — is 2.1 (the extra one-tenth compensates for infant and child mortality). When the number of children per woman stays above 2.1 even slightly, the population still expands exponentially. This means that although the population climbs less and less steeply as the number approaches 2.1, humanity will still, in theory, eventually come to weigh as much as Earth and, if given enough time, will exceed the mass of the visible universe. This fantasy is a mathematician's way of saying that anything above zero population growth cannot be sustained. If, on the other hand, the average number of children drops below 2.1, the population enters negative exponential growth and starts to decline. To speak of 2.1 in exact terms as the breakpoint is of course an oversimplification. Advances in medicine and public health can lower the breakpoint toward the minimal, perfect number of 2.0 (no infant or childhood deaths), while famine, epidemics, and war, by boosting mortality, can raise it well above 2.1. But worldwide, over an extended period of time, local differences and statistical fluctuations wash one another out and the iron demographic laws grind on. They transmit to us always the same essential message, that to breed in excess is to overload the planet.

By 2000 the replacement rate in all of the countries of western Europe had dropped below 2.1. The lead was taken by Italy, at 1.2 children per woman (so much for the power of natalist religious doctrine). Thailand also passed the magic number, as well as the nonimmigrant population of the United States.

When a country descends to its zero-population birth rates or even well below, it does not cease absolute population growth immediately, because the positive growth experienced just before the breakpoint has generated a disproportionate number of young people with most of their fertile years and life ahead of them. As this cohort ages, the proportion of child-bearing people diminishes, the age distribution stabilizes at the zero-population level, the slack is taken up, and population growth ceases. Similarly, when a country dips below the breakpoint, a lag period intervenes before the absolute growth rate goes negative and the population actually declines. Italy and Germany, for example, have entered a period of such true, absolute negative population growth.

The decline in global population growth is attributable to three interlocking social forces: the globalization of an economy driven by science and technology, the consequent implosion of rural populations into cities, and, as a result of globalization and urban implosion, the empowerment of women. The freeing of women socially and economically results in fewer children. Reduced reproduction by female choice can be thought a fortunate, indeed almost miraculous, gift of human nature to future generations. It could have gone the other way: women, more prosperous and less shackled, could have chosen the satisfactions of a larger brood.

They did the opposite. They opted for a smaller number of quality children, who can be raised with better health and education, over a larger family. They simultaneously chose better, more secure lives for themselves. The tendency appears to be very widespread, if not universal. Its importance cannot be overstated. Social commentators often remark that humanity is endangered by its own instincts, such as tribalism, aggression, and personal greed. Demographers of the future will, I believe, point out that on the other hand humanity was saved by this one quirk in the maternal instinct.

The global trend toward smaller families, if it continues, will eventually halt population growth and afterward reverse it. What will be the peak, and when will it occur? And how will the environment fare as humanity climbs to the peak? The Population Division of the United Nations Department of Economic and Social Affairs released a spread of projections to the year 2050 that ranged from 7.3 billion to 14.4 billion, with the most likely scenario falling somewhere between 9 billion and 10 billion.

Enough slack still exists in the system to justify guarded optimism. Women given a choice and affordable contraceptive methods generally practice birth control. By 1996 about 130 countries subsidized family planning services. More than half of all developing countries in particular also had official population policies to accompany their economic and military policies, and more than 90 percent of the rest stated their intention to follow suit. The United States, where the idea is still virtually taboo, remained a stunning exception.

The encouragement of population control by developing countries comes not a moment too soon. The environmental fate of the world lies ultimately in their hands. They now account for virtually all global population growth, and their drive toward higher per capita consumption will be relentless.

The consequences of their reproductive prowess are multiple and deep. The people of the developing countries are already far younger than those in the industrial countries and destined to become more so. The streets of Lagos, Manaus, Karachi, and other cities in the developing world are a sea of children. To an observer fresh from Europe or North America, the crowds give the feel of a gigantic school just let out. In at least sixty-eight of the countries, more than 40 percent of the population is under fifteen years of age.

A country poor to start with and composed largely of young children and adolescents is strained to provide even minimal health services and education for its people. Its superabundance of cheap, unskilled labor can be turned to some economic advantage but unfortunately also provides cannon fodder for ethnic strife and war. As the populations continue to explode and water and arable land grow scarcer, the industrial countries will feel their pressure in the form of many more desperate immigrants and the risk of spreading international terrorism. I have come to understand the advice given me many years ago when I argued the case for the natural environment to the president's scientific adviser: your patron is foreign policy.

Stretched to the limit of its capacity, how many people can the planet support? A rough answer is possible, but it is a sliding one contingent on three conditions: how far into the future the planetary support is expected to last, how evenly the resources are to be distributed, and the quality of life most of humanity expects to achieve. Consider food, which economists commonly use as a proxy of carrying capacity. The current world production of grains, which provide most of humanity's calories, is about 2 billion tons annually. That is enough, in theory, to feed 10 billion East Indians, who eat primarily grains and very little meat by Western standards. But the same amount can support only about 2.5 billion Americans, who convert a large part of their grains into livestock and poultry. There are two ways to stop short of the wall. Either the industrial populations move down the food chain to a more vegetarian diet, or the agricultural yield of productive land worldwide is increased by more than 50 percent.

The constraints of the biosphere are fixed. The bottleneck through which we are passing is real. It should be obvious to anyone not in a euphoric delirium that whatever humanity does or does not do, Earth's capacity to support our species is approaching the limit. We already appropriate by some means or other 40 percent of the planet's organic matter produced by green plants. If everyone agreed to become vegetarian, leaving little or nothing for livestock, the present 1.4 billion hectares of arable land (3.5 billion acres) would support about 10 billion people. If humans utilized as food all of the energy captured by plant photosynthesis on land and sea, some 40 trillion watts, the planet could support about 16 billion people. But long before that ultimate limit was approached, the planet would surely have become a hellish place to exist. There may, of course, be escape hatches. Petroleum reserves might be converted into food until they are exhausted. Fusion energy could conceivably be used to create light, whose energy would power photosynthesis, ramp up plant growth beyond that dependent on solar energy, and hence create more food. Humanity might even consider becoming someday what the astrobiologists call a type II civilization and harness all the power of the sun to support human life on Earth and on colonies on and around the other solar planets. Surely these are not frontiers we will wish to explore in order simply to continue our reproductive folly.

The epicenter of environmental change, the paradigm of population stress, is the People's Republic of China. By 2000 its population was 1.2 billion, one-fifth of the world total. It is thought likely by demographers to creep up to 1.6 billion by 2030. During 1950–2000 China's people grew by 700 million, more than existed in the entire world at the start of the industrial revolution. The great bulk of this increase is crammed into the basins of the Yangtze and Yellow rivers, covering an area about equal to that of the eastern United States. Hemmed in to the west by deserts and mountains, limited to the south by resistance from other civilizations, their agricultural populations simply grew denser on the land their ancestors had farmed for millennia. China became in effect a great overcrowded island, a Jamaica or Haiti writ large.

Highly intelligent and innovative, its people have made the most of it. Today China and the United States are the two leading grain producers of the world. But China's huge population is on the verge of consuming more than it can produce. In 1997 a team of scientists, reporting to the U.S. National Intelligence Council (NIC), predicted that China will need to import 175 million tons of grain annually by 2025. Extrapolated to 2030, the annual level is 200 million tons — the entire amount of grain exported annually in the world at the present time. A tick in the parameters of the model could move these figures up or down, but optimism would be a dangerous attitude in planning strategy when the stakes are so high. After 1997 the Chinese in fact instituted a province-level crash program to boost grain level to export capacity. The effort was successful but may be short-lived, a fact the government itself recognizes. It requires the cultivation of marginal land, higher per-acre environmental damage, and a more rapid depletion of the country's precious groundwater.

According to the NIC report, any slack in China's production may be picked up by the Big Five grain exporters: the United States, Canada, Argentina, Australia, and the European Union. But the exports of these dominant producers, after climbing steeply in the 1960s and 1970s, tapered off to near their present level in 1980. With existing agricultural capacity and technology, this output does not seem likely to increase to any significant degree. The United States and the European Union have already returned to production all of the cropland idled under earlier farm commodity programs. Australia and Canada, largely dependent on dryland farming, are constrained by low rainfall. Argentina has the potential to expand, but due to its small size, the surplus it produces is unlikely to exceed 10 million tons of grain production per year.

China relies heavily on irrigation, with water drawn from its aquifers and great rivers. The greatest impediment is again geographic: two-thirds of China's agriculture is in the north, but four-fifths of the water supply is in the south — that is, principally in the Yangtze River Basin. Irrigation and withdrawals for domestic and industrial use have depleted the northern basins, from which flow the waters of the Yellow, Hai, Huai, and Liao rivers. Starting in 1972, the Yellow River Channel has gone bone dry almost yearly through part of its course in Shandong Province, as far inland as the capital, Jinan, thence down all the way to the sea. In 1997 the river stopped flowing for 130 days, then restarted and stopped again through the year for a record total of 226 dry days. Because Shandong Province normally produces a fifth of China's wheat and a seventh of its corn, the failure of the Yellow River is of no little consequence. The crop losses in 1997 alone reached $1.7 billion.

Meanwhile, the groundwater of the northern plains has dropped precipitously, reaching an average rate of 1.5 meters (5 feet) per year by the mid-1990s. Between 1965 and 1995 the water table fell 37 meters (121 feet) beneath Beijing itself.

Faced with chronic water shortages in the Yellow River Basin, the Chinese government has undertaken the building of the Xiaolangdi Dam, which will be exceeded in size only by the Three Gorges Dam on the Yangtze River. The Xiaolangdi is expected to solve the problems of both periodic flooding and drought. Plans are being laid in addition for the construction of canals to siphon water from the Yangtze, which never grows dry, to the Yellow River and Beijing, respectively.

These measures may or may not suffice to maintain Chinese agriculture and economic growth. But they are complicated by formidable side effects. Foremost is silting from the upriver loess plains, which makes the Yellow River the most turbid in the world and threatens to fill the Xiaolangdi Reservoir, according to one study, as soon as thirty years after its completion.

China has maneuvered itself into a position that forces it continually to design and redesign its lowland territories as one gigantic hydraulic system. But this is not the fundamental problem. The fundamental problem is that China has too many people. In addition, its people are admirably industrious and fiercely upwardly mobile. As a result, their water requirements, already oppressively high, are rising steeply. By 2030 residential demands alone are projected to increase more than fourfold, to 134 billion tons, and industrial demands fivefold, to 269 billion tons. The effects will be direct and powerful. Of China's 617 cities, 300 already face water shortages.

The pressure on agriculture is intensified in China by a dilemma shared in varying degrees by every country. As industrialization proceeds, per capita income rises, and the populace consumes more food. They also migrate up the energy pyramid to meat and dairy products. Because fewer calories per kilogram of grain are obtained when first passed through poultry and livestock instead of being eaten directly, per capita grain consumption rises still more. All the while the available water supply remains static or nearly so. In an open market, the agricultural use of water is outcompeted by industrial use. A thousand tons of fresh water yields a ton of wheat, worth $200, but the same amount of water in industry yields $14,000. As China, already short on water and arable land, grows more prosperous through industrialization and trade, water becomes more expensive. The cost of agriculture rises correspondingly, and unless the collection of water is subsidized, the price of food also rises. This is in part the rationale for the great dams at Three Gorges and Xiaolangdi, built at enormous public expense.

In theory, an affluent industrial country does not have to be agriculturally independent. In theory, China can make up its grain shortage by purchasing from the Big Five grain-surplus nations. Unfortunately, its population is too large and the world surplus too restrictive for it to solve its problem without altering the world market. All by itself, China seems destined to drive up the price of grain and make it harder for the poorer developing countries to meet their own needs. At the present time, grain prices are falling, but this seems certain to change as the world population soars to 9 billion or beyond.

The problem, resource experts agree, cannot be solved entirely by hydrological engineering. It must include shifts from grain to fruit and vegetables, which are more labor-intensive, giving China a competitive edge. To this can be added strict water conservation measures in industrial and domestic use; the use of sprinkler and drip irrigation in cultivation, as opposed to the traditional and more wasteful methods of flood and furrow irrigation; and private land ownership, with subsidies and price liberalization, to increase conservation incentives for farmers.

Meanwhile the surtax levied on the environment to support China's growth, though rarely entered on the national balance sheets, is escalating to a ruinous level. Among the most telling indicators is the pollution of water. Here is a measure worth pondering. China has in all 50,000 kilometers of major rivers. Of these, according to the U.N. Food and Agriculture Organization, 80 percent no longer support fish. The Yellow River is dead along much of its course, so fouled with chromium, cadmium, and other toxins from oil refineries, paper mills, and chemical plants as to be unfit for either human consumption or irrigation. Diseases from bacterial and toxic-waste pollution are epidemic.

China can probably feed itself to at least midcentury, but its own data show that it will be skirting the edge of disaster even as it accelerates its lifesaving shift to industrialization and megahydrological engineering. The extremity of China's condition makes it vulnerable to the wild cards of history. A war, internal political turmoil, extended droughts, or crop disease can kick the economy into a downspin. Its enormous population makes rescue by other countries impracticable.

China deserves close attention, not just as the unsteady giant whose missteps can rock the world, but also because it is so far advanced along the path to which the rest of humanity seems inexorably headed. If China solves its problems, the lessons learned can be applied elsewhere. That includes the United States, whose citizens are working at a furious pace to overpopulate and exhaust their own land and water from sea to shining sea.

Environmentalism is still widely viewed, especially in the United States, as a special interest lobby. Its proponents, in this blinkered view, flutter their hands over pollution and threatened species, exaggerate their case, and press for industrial restraint and the protection of wild places, even at the cost of economic development and jobs.

Environmentalism is something more central and vastly more important. Its essence has been defined by science in the following way. Earth, unlike the other solar planets, is not in physical equilibrium. It depends on its living shell to create the special conditions on which life is sustainable. The soil, water, and atmosphere of its surface have evolved over hundreds of millions of years to their present condition by the activity of the biosphere, a stupendously complex layer of living creatures whose activities are locked together in precise but tenuous global cycles of energy and transformed organic matter. The biosphere creates our special world anew every day, every minute, and holds it in a unique, shimmering physical disequilibrium. On that disequilibrium the human species is in total thrall. When we alter the biosphere in any direction, we move the environment away from the delicate dance of biology. When we destroy ecosystems and extinguish species, we degrade the greatest heritage this planet has to offer and thereby threaten our own existence.

Humanity did not descend as angelic beings into this world. Nor are we aliens who colonized Earth. We evolved here, one among many species, across millions of years, and exist as one organic miracle linked to others. The natural environment we treat with such unnecessary ignorance and recklessness was our cradle and nursery, our school, and remains our one and only home. To its special conditions we are intimately adapted in every one of the bodily fibers and biochemical transactions that give us life.

That is the essence of environmentalism. It is the guiding principle of those devoted to the health of the planet. But it is not yet a general worldview, evidently not yet compelling enough to distract many people away from the primal diversions of sport, politics, religion, and private wealth.

The relative indifference to the environment springs, I believe, from deep within human nature. The human brain evidently evolved to commit itself emotionally only to a small piece of geography, a limited band of kinsmen, and two or three generations into the future. To look neither far ahead nor far afield is elemental in a Darwinian sense. We are innately inclined to ignore any distant possibility not yet requiring examination. It is, people say, just good common sense. Why do they think in this shortsighted way? The reason is simple: it is a hardwired part of our Paleolithic heritage. For hundreds of millennia, those who worked for short-term gain within a small circle of relatives and friends lived longer and left more offspring — even when their collective striving caused their chiefdoms and empires to crumble around them. The long view that might have saved their distant descendants required a vision and extended altruism instinctively difficult to marshal.

The great dilemma of environmental reasoning stems from this conflict between short-term and long-term values. To select values for the near future of one's own tribe or country is relatively easy. To select values for the distant future of the whole planet also is relatively easy — in theory, at least. To combine the two visions to create a universal environmental ethic is, on the other hand, very difficult. But combine them we must, because a universal environmental ethic is the only guide by which humanity and the rest of life can be safely conducted through the bottleneck into which our species has foolishly blundered.

Contributors' Notes

*Other Notable Science and
Nature Writing of 2002*

Contributors' Notes

Natalie Angier is a Pulitzer Prize–winning journalist and the author of the national bestseller *Woman: An Intimate Geography*. Her two previous works, *The Beauty of the Beastly* and *Natural Obsessions*, were named New York Times Notable Books. In addition, Angier writes for the *New York Times, Time,* and *Discover.* She lives with her family in Takoma Park, Maryland.

Tim Appenzeller is the science and technology editor at *U.S. News & World Report.* For more than twenty years, he has been an editor at publications that include *Scientific American, The Sciences,* and the news section of *Science,* but he writes whenever he can.

Alan Burdick is a former editor for the *New York Times Magazine* and *The Sciences.* His writing has appeared in those magazines and many others, including *Harper's, GQ, Wired,* and *Outside;* he is a regular contributor to *Discover* and *Natural History.* He worked most recently as senior writer and producer for *Science Bulletins* at the American Museum of Natural History. He is currently completing a book about nature.

Clark R. Chapman is a planetary scientist at the Boulder, Colorado, office of Southwest Research Institute. He has served on the imaging and spectroscopy Science Teams of the Galileo mission to Jupiter, the NEAR Shoemaker mission to the near-Earth asteroid Eros, and the forthcoming MESSENGER mission to orbit the planet Mercury. Dr. Chapman has written hundreds of technical articles as well as many articles and books for lay readers and is a coauthor of *Cosmic Catastrophes.* He is the 1999 recipient of the Carl Sagan Medal for Excellence in Public Communication of Planetary Science. The asteroid 2409 Chapman is named for him.

David Ewing Duncan is a contributing editor to *Wired;* he has written for *The Atlantic Monthly, Harper's Magazine, Discover,* and *Smithsonian* and was a longtime correspondent for *Life.* Duncan is the author of the best-selling *Calendar: Humanity's Epic Struggle to Determine a True and Accurate Year,* as well as *Residents: The Perils and Promise of Educating Young Doctors; Hernando de Soto: A Savage Quest in the Americas; Cape to Cairo;* and *Pedaling the Ends of the Earth.* A commentator and guest producer on NPR's *Morning Edition,* Duncan was a special producer for ABC's *Nightline* and *20/20* and a producer for Discovery Television. He lives in San Francisco and is working on two books, *Masterminds of Biotech* and *Copernicus's Monster.* His Web site is www.literati.net/Duncan.

Timothy Ferris is the author of ten books, among them the bestsellers *The Whole Shebang* and *Coming of Age in the Milky Way,* which have been translated into fifteen languages and were named by the *New York Times* as two of the leading books published in the twentieth century. He is also the editor of two anthologies, *The Best American Science Writing 2001* and *The World Treasury of Physics, Astronomy, and Mathematics.* A former newspaper reporter and editor of *Rolling Stone,* Ferris is a frequent contributor to *The New Yorker* and *The New York Review of Books.*

Ian Frazier, a staff writer at *The New Yorker* for twenty-one years, is the author of *On the Rez, Great Plains, Coyote v. Acme,* and *Dating Your Mom.* He divides his time between homes in New Jersey and Montana.

James Gorman is a science reporter for the *New York Times.* He has been writing about science for almost thirty years and has produced several books, including *Digging Dinosaurs,* with John R. Horner, and *The Man with No Endorphins,* a collection of columns he wrote for *Discover* magazine in the 1980s.

Alan W. Harris specializes in theoretical and observational research dealing with the orbital and rotational dynamics of planets and the smaller bodies of the solar system. Recently Dr. Harris has advised various U.S. and foreign government commissions on the hazard from wayward Near-Earth Objects (asteroids and comets), which could slam into Earth with devastating consequences. Dr. Harris also maintains an active interest in debunking pseudoscience and paranormal claims and recently participated in the Fourth World Skeptics Conference to discuss a case of "pathological astronomical science" involving largely nonexistent tiny comets.

Charles Hirshberg hails from a distinguished family of scientists, but he missed the smarty-pants gene and so became a journalist. He began at the

Washington Post Sunday Magazine, then moved to the *Los Angeles Times* and, for ten years, served as a staff writer for *Life* magazine. The author of three books on popular music, he is currently at work on a book on the social implications of animal abuse.

Brendan I. Koerner is a contributing editor at *Wired* and a fellow at the New America Foundation. He was formerly a senior editor at *U.S. News & World Report,* where he covered everything from astrophysics to Vatican politics. His work has appeared in the *New York Times Magazine, Harper's Magazine, Slate, Mother Jones, Washington Monthly, The New Republic,* and *Legal Affairs.* He also writes the "Mr. Roboto" technology column for the *Village Voice.* Koerner was named one of the *Columbia Journalism Review*'s "Ten Young Writers on the Rise" in 2002, and he recently won a National Headliner Award for feature writing. A 1996 graduate of Yale, he lives in New York City.

Elizabeth Kolbert has been a staff writer for *The New Yorker* since 1999. Earlier she held a number of positions at the *New York Times,* including "Metro Matters" columnist and Albany bureau chief. Ms. Kolbert, a graduate of Yale University, lives in Westchester County with her husband and their three sons.

Andrew Lawler is the Boston correspondent for *Science* magazine, covering space, archaeology, and everything in between. In 1999 he was an MIT Knight Science Journalism fellow, after seventeen years of reporting on space and science issues in Washington, D.C. He has written for *Smithsonian, Air & Space Magazine,* and *Discover.*

Daniel Lazare is a writer and journalist who has written about politics and constitutional theory for a wide variety of publications, including *Harper's Magazine, The Nation, New Left Review,* and *Le Monde Diplomatique.* His books include *The Frozen Republic,* an iconoclastic study of the U.S. Constitution; *America's Undeclared War,* an examination of U.S. urban policy; and *The Velvet Coup,* a constitutional analysis of the Bush-Gore electoral crisis. He is currently at work on a book about monotheism and political power. He lives in Manhattan.

Elizabeth F. Loftus is Distinguished Professor at the University of California at Irvine. Her book *Eyewitness Testimony* won a National Media Award (Distinguished Contribution) from the American Psychological Foundation, and *The Myth of Repressed Memory* (coauthored with Katherine Ketcham) has been translated into many languages. In 2002 she was named to the list of the one hundred most eminent psychologists of the twentieth century and was the top-ranked woman.

Charles C. Mann's most recent book is *@ Large,* the true story of a bizarre, calamitous episode in the history of computer security and the Internet. A correspondent for *The Atlantic Monthly* and *Science,* he has covered the intersection of science, technology, and commerce for many newspapers and magazines here and abroad. His next book is an account of new research on the pre-Columbian Americas; an excerpt, "1491," was published in *The Atlantic Monthly* in 2002.

Bill McKibben is best known as the author of *The End of Nature,* the first book for a general audience about global warming; it has been translated into twenty languages. A former staff writer for *The New Yorker,* he writes regularly for many publications, including *The Atlantic Monthly, Harper's Magazine,* and *The New York Review of Books.* His most recent book is *Enough: Staying Human in an Engineered Age.* He is now a scholar in residence at Middlebury College.

Steve Olson is the author of *Mapping Human History: Genes, Race, and Our Common Origins,* a finalist for the 2002 nonfiction National Book Award. He also has written articles for *The Atlantic Monthly, Science,* the *Washington Post, Slate, Teacher, Astronomy,* and other magazines. His current book is a narrative account of the 2001 International Mathematical Olympiad. He lives near Washington, D.C., with his wife and two children.

Dennis Overbye is a science reporter for the *New York Times.* He is the author of *Einstein in Love: A Scientific Romance,* which was a Los Angeles Times Book Prize finalist, and *Lonely Hearts of the Cosmos: The Story of the Scientific Quest for the Secret of the Universe,* a finalist for the National Book Critics Circle Award and the Los Angeles Times Book Prize and the winner of the American Institute of Physics award for science writing. He lives in Manhattan with his wife, Nancy Wartik, and their daughter.

Steven Pinker is the author of critically acclaimed popular science books, including *The Language Instinct* and *How the Mind Works,* a finalist for the Pulitzer Prize in 1998. He is an elected fellow of several scholarly societies, including the American Academy of Arts and Sciences, the American Association for the Advancement of Science, and the Neuroscience Research Program. Professor Pinker writes frequently in the popular press, including the *New York Times, Time, The New Yorker,* and *Technology Review.* His sixth book, *The Blank Slate,* was published in the fall of 2002.

Oliver Sacks was born in London and educated in London, Oxford, and California. A practicing neurologist, he has written nine books, including *The Man Who Mistook His Wife for a Hat* and *Awakenings*. His work frequently appears in *The New Yorker* and *The New York Review of Books*, among other publications, and has won numerous awards, including a George S. Polk Award, a Guggenheim Fellowship, and an Alfred P. Sloan Foundation grant. His most recent books are *Uncle Tungsten: Memories of a Chemical Boyhood* and *Oaxaca Journal*.

Steve Silberman is a contributing editor of *Wired*, where he has written about autism in Silicon Valley, the FBI, the future of surveillance, "quorum sensing" in bacteria, science and religion, computer translation, and the *Matrix* sequels. His writing has appeared in *The New Yorker, Time,* Salon. com, *Tikkun,* and many other publications. He also coproduced the Grateful Dead's *So Many Roads* (*1965–1995*). He lives in San Francisco, and his home page on the Web is www.levity.com/digaland.

Adam Summers grew up in New York City and the north woods of Canada and showed an early proclivity for taking mechanical systems to pieces. At Swarthmore College he learned to put them back together again by taking degrees in engineering and mathematics. He studied biology at the University of Massachusetts and spent an inordinate amount of time wandering among the collections of pickled fish, amphibians, and reptiles at the Museum of Comparative Zoology at Harvard. A postdoctoral sojourn to Berkeley revealed the excellent intellectual and physical climate of the West Coast, and he has since settled at the University of California at Irvine. Summers has published over thirty scientific articles and has contributed a monthly column to *Natural History* for the past two years.

Gary Taubes has written about science, medicine, and health for *Science, Discover, The Atlantic Monthly,* the *New York Times Magazine,* and a host of other publications. He is currently a contributing correspondent for *Science* and a contributing editor for *Technology Review.* He has won numerous awards for his reporting, including the National Association of Science Writers Science-in-Society Journalism Award in 1996, 1999, and 2001. His most recent book, *Bad Science: The Short Life and Weird Times of Cold Fusion,* was a New York Times Notable Book and a finalist for the Los Angeles Times Book Award. He is working now on a book about diet, obesity, and chronic disease, to be published in January 2005.

Bruce Watson is the author of *The Man Who Changed How Boys and Toys Were Made,* which was cited by the *Christian Science Monitor* among its best nonfiction books of 2002. He is a frequent contributor to *Smithsonian,* writing on subjects ranging from eels to Ferraris to a history of Coney Island. His articles have also appeared in the *Los Angeles Times, Reader's Digest,* the *Boston Globe, Yankee,* and other publications. He lives in western Massachusetts.

William Speed Weed has written for *Discover, National Geographic Adventure,* the *New York Times Magazine, Popular Science, Astronomy,* and Salon.com, among others. He lives in Syracuse, New York, with his girlfriend and two puppies. Speed is indeed his real name.

Scott Weidensaul is the author of more than two dozen books on natural history, including *Mountains of the Heart: A Natural History of the Appalachians,* the Pulitzer Prize–nominated *Living on the Wind: Across the Hemisphere with Migratory Birds,* and his latest book, *The Ghost With Trembling Wings: Science, Wishful Thinking, and the Search for Lost Species.* Weidensaul writes regularly for *Smithsonian,* and his work has appeared in such publications as *Audubon,* the *New York Times, Natural History, International Wildlife,* and *Orion.*

Steven Weinberg is a professor at the University of Texas at Austin, where he founded its Theory Group. He has been honored with numerous awards, including the Nobel Prize in physics and the National Medal of Science. In addition to the treatises *Gravitation and Cosmology* and *The Quantum Theory of Fields,* Weinberg has written several books, including the prizewinning *The First Three Minutes, The Discovery of Subatomic Particles, Dreams of a Final Theory,* and, most recently, *Facing Up: Science and Its Cultural Adversaries.* He is a frequent contributor to *The New York Review of Books.*

Ted Williams has been writing full-time on environmental issues, with special attention to fish and wildlife conservation, since 1970. In addition to freelancing for national magazines, he contributes regularly to *Audubon* and *Fly Rod and Reel.* In 2003 the Federation of Fly Fishers presented him with its Aldo Leopold Award for "outstanding contributions to fisheries and land ecology," and "Maine's War on Coyotes" was named a National Magazine Award finalist.

Edward O. Wilson is the author of two Pulitzer Prize–winning books, *On Human Nature* and *The Ants.* His most recent book is *The Future of Life.* He has received many fellowships, honors, and awards, including the 1976 National Medal of Science. He lives with his wife in Lexington, Massachusetts.

Other Notable Science and Nature Writing of 2002

SELECTED BY TIM FOLGER

DANA MACKENZIE
The Stanford Flip. *Discover,* October.
CHARLES C. MANN
1491. *The Atlantic Monthly,* March.
PETER MATTHIESSEN
Burning Bright. *Outside,* October.
DOUGLAS McGRAY
Biotech's Black Market. *Mother Jones,* September/October.
BEN MEZRICH
Hacking Las Vegas. *Wired,* September.
STEVE MIRSKY
Members Only. *Scientific American,* December.
Poultry and Poetry. *Scientific American,* August.
MARK W. MOFFETT
Bit. *Outside,* April.
MICHAEL MOYER
Antimatter. *Popular Science,* April.

J. MADELEINE NASH
The Secrets of Autism. *Time,* May 6.

DENNIS OVERBYE
Peering Through the Gates of Time. *The New York Times,* March 12.

JAKE PAGE
Dragonfly Dramas. *Smithsonian,* January.
RICHARD PANEK
And Then There Was Light. *Natural History,* November.
MICHAEL PARFIT
Lost at Sea. *Smithsonian,* April.

EVAN RATLIFF
This Is Not a Test. *Wired,* March.
JONATHAN RAUCH
Seeing Around Corners. *The Atlantic Monthly,* April.

ROBERT M. SAPOLSKY
Cheaters and Chumps. *Natural History,* June.
JOHN SEIDENSTICKER
Tiger Tracks. *Smithsonian,* January.
BRUCE SELCRAIG
Digging Ditches. *Smithsonian,* February.
HAMPTON SIDES
Crawl Space. *Outside,* February.
NATASHA SINGER
These Pants Saved My Life. *Outside,* October.

FLOYD SKLOOT
  The Melody Lingers On. *Southwest Review,* Vol. 87, Nos. 2 and 3.
ANNICK SMITH
  Sacred Ground. *Audubon,* January/February.
TOM STANDAGE
  Monster in a Box. *Wired,* March.
FRED STREBEIGH
  Defending Russian Nature. *Sierra,* March/April.

SALLIE TISDALE
  Seeing Clearly. *Audubon,* January/February.
NEIL DE GRASSE TYSON
  Delusions of Centrality. *Natural History,* December.

ELLEN ULLMAN
  Programming the Post-Human. *Harper's Magazine,* October.

WILLIAM T. VOLLMANN
  Salton Sea. *Outside,* February.

JACOB WARD
  Crime Seen. *Wired,* May.
MARK WHEELER
  California Scheming. *Smithsonian,* October.
TED WILLIAMS
  A Crossroad for Wilderness. *Mother Jones,* September/October.
KAREN WRIGHT
  Six Degrees of Speculation. *Discover,* June.

MATTHEW YEOMANS
  Unplugged. *Wired,* February.

THE B·E·S·T AMERICAN SERIES ™

THE BEST AMERICAN SHORT STORIES® 2003
Walter Mosley, guest editor • Katrina Kenison, series editor

"Story for story, readers can't beat the *Best American Short Stories* series" (*Chicago Tribune*). This year's most beloved short fiction anthology is edited by the award-winning author Walter Mosley and includes stories by Dorothy Allison, Mona Simpson, Anthony Doerr, Dan Chaon, and Louise Erdrich, among others.

0-618-19733-8 PA $13.00 / 0-618-19732-X CL $27.50
0-618-19748-6 CASS $26.00 / 0-618-19752-4 CD $35.00

THE BEST AMERICAN ESSAYS® 2003
Anne Fadiman, guest editor • Robert Atwan, series editor

Since 1986, the *Best American Essays* series has gathered the best non-fiction writing of the year and established itself as the best anthology of its kind. Edited by Anne Fadiman, author of *Ex Libris* and editor of the *American Scholar*, this year's volume features writing by Edward Hoagland, Adam Gopnik, Michael Pollan, Susan Sontag, John Edgar Wideman, and others.

0-618-34161-7 PA $13.00 / 0-618-34160-9 CL $27.50

THE BEST AMERICAN MYSTERY STORIES™ 2003
Michael Connelly, guest editor • Otto Penzler, series editor

Our perennially popular anthology is a favorite of mystery buffs and general readers alike. This year's volume is edited by the best-selling author Michael Connelly and offers pieces by Elmore Leonard, Joyce Carol Oates, Brendan DuBois, Walter Mosley, and others.

0-618-32965-X PA $13.00 / 0-618-32966-8 CL $27.50
0-618-39072-3 CD $35.00

THE BEST AMERICAN SPORTS WRITING™ 2003
Buzz Bissinger, guest editor • Glenn Stout, series editor

This series has garnered wide acclaim for its stellar sports writing and top-notch editors. Now Buzz Bissinger, the Pulitzer Prize–winning journalist and author of the classic *Friday Night Lights,* continues that tradition with pieces by Mark Kram Jr., Elizabeth Gilbert, Bill Plaschke, S. L. Price, and others.

0-618-25132-4 PA $13.00 / 0-618-25130-8 CL $27.50

THE B·E·S·T AMERICAN SERIES

THE BEST AMERICAN TRAVEL WRITING 2003
Ian Frazier, guest editor • Jason Wilson, series editor

The Best American Travel Writing 2003 is edited by Ian Frazier, the author of *Great Plains* and *On the Rez*. Giving new life to armchair travel this year are William T. Vollmann, Geoff Dyer, Christopher Hitchens, and many others.

0-618-11881-0 PA $13.00 / 0-618-11881-0 CL $27.50
0-618-39074-X CD $35.00

THE BEST AMERICAN SCIENCE AND NATURE WRITING 2003
Richard Dawkins, guest editor • Tim Folger, series editor

This year's edition promises to be another "eclectic, provocative collection" (*Entertainment Weekly*). Edited by Richard Dawkins, the eminent scientist and distinguished author, it features work by Bill McKibben, Steve Olson, Natalie Angier, Steven Pinker, Oliver Sacks, and others.

0-618-17892-9 PA $13.00 / 0-618-17891-0 CL $27.50

THE BEST AMERICAN RECIPES 2003–2004
Edited by Fran McCullough and Molly Stevens

"The cream of the crop . . . McCullough's selections form an eclectic, unfussy mix" (*People*). Offering the very best of what America is cooking, as well as the latest trends, time-saving tips, and techniques, this year's edition includes a foreword by Alan Richman, award-winning columnist for *GQ*.

0-618-27384-0 CL $26.00

THE BEST AMERICAN NONREQUIRED READING 2003
Edited by Dave Eggers • Introduction by Zadie Smith

Edited by Dave Eggers, the author of *A Heartbreaking Work of Staggering Genius* and *You Shall Know Our Velocity*, this genre-busting volume draws the finest, most interesting, and least expected fiction, nonfiction, humor, alternative comics, and more from publications large, small, and on-line. *The Best American Nonrequired Reading 2003* features writing by David Sedaris, ZZ Packer, Jonathan Safran Foer, Andrea Lee, and others.

0-618-24696-7 $13.00 PA / 0-618-24696-7 $27.50 CL
0-618-39073-1 $35.00 CD

HOUGHTON MIFFLIN COMPANY   www.houghtonmifflinbooks.com